REMEMBERING THE GIANTS
APOLLO ROCKET PROPULSION DEVELOPMENT

Editors:

Steven C. Fisher

Shamim A. Rahman

John C. Stennis Space Center

The NASA History Series

National Aeronautics and Space Administration

NASA History Division
Office of External Relations
Washington, DC
December 2009
NASA SP-2009-4545

Table of Contents

Foreword 7

Acknowledgments 9

Welcome Remarks
     Richard Gilbrech 11
     Steve Fisher...13

Chapter One - Robert Biggs, Rocketdyne - F-1 Saturn V First Stage Engine 15

Chapter Two - Paul Coffman, Rocketdyne - J-2 Saturn V 2nd & 3rd Stage Engine.............27

Chapter Three - Gerald R. Pfeifer, Aerojet - Attitude Control Engines 39

Chapter Four - Tim Harmon, Rocketdyne - SE-7 & SE-8 Engines 51

Chapter Five - Clay Boyce, Aerojet - AJ10-137 Apollo Service Module Engine 59

Chapter Six - Gerard Elverum, TRW - Lunar Descent Engine..73

Chapter Seven - Tim Harmon, Rocketdyne - Lunar Ascent Engine 87

Appendix A - Event Program 97

Appendix B - Speakers' Group Photograph 101

Appendix C - Robert Biggs' Presentation Viewgraphs...103

Appendix D - Paul Coffman's Presentation Viewgraphs 113

Appendix E - G. R. "Jerry" Pfeifer's Presentation Viewgraphs..123

Appendix F - Tim Harmon's SE-7 & SE-8 Presentation Viewgraphs 133

Appendix G - Clay Boyce's Presentation Viewgraphs 143

Appendix H - Jerry Elverum's Presentation Viewgraphs..151

Appendix I - Tim Harmon's Lunar Ascent Engine Presentation Viewgraphs.....................171

Appendix J - Event Photos and Highlights 179

Further Reading 191

Glossary 193

The NASA History Series 197

On April 25, 2006, NASA's John C. Stennis Space Center hosted a series of lectures on Apollo Propulsion development. This monograph is a transcript of the event, held as part of the celebration to mark the 40th anniversary of the first rocket engine test conducted at the site then known as the Mississippi Test Facility. On April 23, 1966, engineers tested a cluster of five J-2 engines that powered the second stage of the Saturn V moon rocket.

This transcript has been edited for readability and clarity. The opinions expressed are solely those of the individuals presented. The report does not in any way promulgate policies or state the official opinions of the National Aeronautics and Space Administration or the U.S. government.

John C. Stennis Space Center
History Office
B-1100
Stennis Space Center, MS 39529
228.688.2646

2006 Event Title *"On the Shoulders of Giants," Apollo Propulsion Development Seminars.*
Moderator: Steve Fisher, Rocketdyne
Speakers: Robert Biggs, Paul Coffman, Gerald Pfeifer, Clay Boyce, Gerard Elverum, and Tim Harmon
Initial Transcription: Michele Beisler, Transcription/Technical Writing: B. Nicole Wells (Jacobs Technology Facility Operating Services Contract)
Editors: Mr. Steve Fisher, P&W Rocketdyne, and Dr. Shamim Rahman, NASA
Event Coordinator: Rebecca Strecker, NASA

About the Cover –
At left, the Apollo Saturn launch vehicle; middle, a J-2 engine on a static test stand operating on oxygen and hydrogen; and at right, lunar surface photographed from approaching spacecraft.

Dr. Shamim Rahman began his professional career at The Aerospace Corporation in El Segundo, CA, in 1985 as an aerospace engineer on launch vehicle and spacecraft flight programs for the Air Force Space and Missile Command. He later joined the TRW Propulsion Research Center to work on innovative rocket propulsion devices for civil and military space applications. Dr. Rahman joined NASA in 1998. At Stennis Space Center (SSC), Dr. Rahman has provided technical leadership and oversight on national rocket propulsion test facilities, and over a variety of research and development test activities for advanced rocket propulsion development. Dr. Rahman assumed his current position in November, 2007, as the Deputy Director for the Engineering & Test Directorate at NASA SSC.

Mr. Steve Fisher has spent 37 years at Rocketdyne in the design, development, testing, and evaluation of liquid rocket engines, and recently retired as a Technical Fellow from Pratt & Whitney Rocketdyne. He initially joined North American Aviation of El Segundo to work on the original B-1 Bomber, and later transferred to the original Rocketdyne Division of the company. During his tenure, he was integrally involved in the development of the Space Shuttle Main Engine components at the Coca 1 and Coca 4 test facilities of Rocketdyne, and has further contributed to a wide variety of propulsion activities including advanced research and development projects with storable and cryogenic propellants, as well as flight engine projects such as the RS-68 engine design and development. Mr. Fisher served as a Boeing Technical Fellow prior to the Rocketdyne division being sold to Pratt & Whitney.

Foreword

It gives us great pleasure to provide this historical compendium of what will likely be remembered as one of the most remarkable achievements in the evolution of rocket propulsion. This achievement was the simultaneous development, testing, and flight use of a series of first-ever propulsive devices that delivered Apollo 11 astronauts safely to the surface of the moon and back to Earth. These devices helped assure three individuals, Armstrong, Aldrin, and Collins a place in the history of humankind.

From the F-1 booster engine to the lunar module ascent engine of the Apollo vehicle stack – all built and delivered by the new United States space industrial base – these individual rocket propulsion development stories provide a glimpse of how technical ingenuity rose to meet the challenge of the race to the moon.

The development histories and lessons learned about the various engines are told by the engineers and project managers, and were recorded on DVD so that the lecture series held at NASA's John C. Stennis Space Center near Bay St. Louis, Mississippi, could be replayed again and thus live on. Remarkably, to those who attended, it was apparent that these speakers recalled their Apollo challenges as if they had happened "just yesterday." It was clear in their voices that the engines carried not just the hardware but also the hope of the nation that this "moon shot" could even be done at all.

Although this monograph comes some years after the actual date of the lectures, and describes work from decades ago, the lessons will continue to carry space exploration forward. The story told within is not how one particular engine was built, but rather how ordinary people persisted and were driven to do extraordinary work. The country owes these resourceful and dedicated engineers a debt of gratitude for giving us the technical precedents upon which today's space programs rest in a continuing story of human exploration.

It would not have been possible without the sanction and enthusiastic support of NASA Stennis Space Center's "front office" (center director, deputy director, and associate director), and the excellent support and facilitation of the local NASA public affairs staff. The 2006 event was officially designated at NASA SSC as "On the Shoulders of Giants", and in this monograph is more aptly designated by the title, "Remembering the Giants."

Shamim A. Rahman, PhD,
NASA Stennis Space Center

Steven C. Fisher,
Technical Fellow, Boeing and
Pratt & Whitney
Retired

Acknowledgments

The initial transcription of these seminars was done by Michele Beisler and Shelia Reed, based upon the excellent videography of Rex Cooksey of CSC who provides Stennis' media services.

This document then was ably constructed by Nicole Wells of Jacobs Technology in draft form to be further edited and taken into publishable form by an excellent media services team. This team included Gene Coleman, Shannon Simpson, Courtney Thomas, and Lacy Thompson.

We wish to recognize Angela Lane and the members of the Media Services Department at the Stennis Space Center for their assistance with graphics and layouts.

This could not have been published without the full support of the NASA History Office at Stennis Space Center, and also NASA's History Division at Headquarters, and in particular Steve Garber who has steered many such documents.

Special thanks also go to Rebecca Strecker, Paul Foerman and Tessa Quave of the NASA Stennis Space Center External Affairs Office, who deployed their excellent support team whenever needed.

Also at NASA Headquarters, thanks go to the professionals in the Communications Support Services Center (CSSC). Printing specialists Tun Hla and Hanta Ralay managed this critical final step. CSSC managers Gail Carter-Kane and Cynthia Miller, as well as Michael Crnkovic and Thomas Powers in the Information Technology and Communications Division, oversaw this project as a whole and made it possible.

Our gratitude to all for their professional and expert contributions.

Panel of speakers seated from left to right: Boyce, Elverum, Harmon, Pfeifer, Coffman and Biggs.

Welcome Remarks

Dr. Richard Gilbrech

We at NASA Stennis Space Center are honored to host *On the Shoulders of Giants*. We have a big mission ahead of us, as did those who worked at the start of the Mercury, Gemini, and Apollo days. We really appreciate the participants coming to Stennis and sharing their wealth of experience with us. Thank you to everyone else attending the event: our friends from Marshall Space Flight Center, Johnson Space Center, and Kennedy Space Center. I have always been in awe of the Mercury, Gemini and Apollo programs. That is really what got me hooked into NASA: the accomplishments that were made in that era, given the state of modeling and computers compared to what we have today. It has always fascinated me. I always felt like I had missed probably one of the most exciting times in engineering history with that feat. My generation got the Space Shuttle Program, which has been exciting, but I always had the "bug" to be part of a "moon shot." I'm very excited that we are getting that chance now. I had the opportunity to accompany Gen. Tom Stafford on a tour of one of the Stennis Space Center test stands. Of course, he had flown two Gemini missions and a Saturn V mission on Apollo 10, and then, later, the Apollo/Soyuz Test Program. I said to him, "I really felt like I missed out." He said, "You are in as exciting a time as I had, and your time will come also." He also commented on the Gemini ride versus the Apollo ride. He said, "The Apollo rides were like a Cadillac compared to the roll maneuvers they did on the Gemini." It is our turn now. I'm very excited.

We do have a big challenge ahead of us. There are tremendous technical challenges, and one thing we don't want to do is repeat mistakes of the past. That is part of what the *On the Shoulders of Giants* program is about: to listen to the wealth of knowledge from these panelists. They have forgotten more about stage testing than I'll ever know. I'm happy to have them participate in this event. Stennis Space Center is poised to play its role in this. We hope to see those clusters of engines hanging on stages and upper stage tests, and then, eventually, to see that cargo lifter we hope will be plugged into the B-2 Stand. We can maybe break some windows again.

One of my mentors, Roy Estess[1], always used to pick on the hardware guys. He'd say, "Stennis has a long history of testing rocket engines and we've even got a longer history of waiting on delivery of rocket engine hardware." I know there is difficulty on the development side, and we appreciate all the challenges that the hardware developers are up against. One thing we want to do is make sure we are poised to get the kind of test data needed to make sure all the test lessons we learned in the Apollo Program are clear and present with our engineers today, so the developers get the data they need to make the decisions they are going to be facing.

---

[1] Roy Estess served as SSC Center Director from 1989 to 2003.

*S-1C Stage Installation into Hot Fire Stand at Mississippi Test Facility (MTF) - 1967
(NASA image number: GPN-2000-000559)*

Steve Fisher

I am really happy to participate in the *On the Shoulders of Giants* series, especially among the six Apollo experts who will speak during the event. It kind of makes me feel like a kid, which is neat. These guys were doing stuff back when I was watching the result on TV. So, this is really exciting for me. It was interesting to pull the lecture series together. It was done before at a Joint Propulsion Conference[2], and we wanted to do it again during summer or fall 2005, but different things prevented that from happening. If you ever had to coordinate with four different companies, four different management groups, and expert people from those four companies, you know what kind of interests that raises. Well, you ought to try doing that with six retired folks. These people are not really retired; these people are active. They are all out doing stuff, and they often tell me, "Hey, we have things to do. We just can't drop what we are doing and show up for a conference." I really want to thank them for participating in this event. It really was a challenge to coordinate and get them all to Stennis Space Center together, and I appreciate them all showing up.

Going to the moon was a monumental undertaking. There's no doubt about that. It required seven different rocket propulsion systems to get the vehicle there and back again. The participants in this lecture series discussed each one of those systems. These were the folks who actually made that dream come true. The series of presentations was very similar to what was shown and what was presented at the 2004 Joint Propulsion Conference during the 35th anniversary of the Apollo 11 first landing on the moon. What was quite interesting to me was that while a lot of these veterans had heard each other's names mentioned and knew of them, some of these folks had never met before that conference. I thought it was quite interesting to see the interaction of these folks as they finally got together after all these years.

---

[2] 2004 Joint Propulsion Conference held at Fort Lauderdale, Florida, sponsored by the American Institute of Aeronautics and Astronautics (AIAA).

F-1 Engine (NASA image number: MSFC-6413912)

Chapter One

Robert Biggs
Rocketdyne - F-1 Saturn V First Stage Engine

Robert "Bob" Biggs worked forty-seven years at Rocketdyne, and spent nine years as lead development engineer and development project engineer on the F-1 Engine Program. He spent several months on the Navaho cruise missile project, three years as lead engineer in the Jupiter Program performance analysis group, and a year as manager of the Dynamic Analysis Laboratory. Biggs also worked thirty-four years on the space shuttle main engine, serving as development manager and chief project engineer.

Before I go into the history of F-1, I want to discuss the F-1 engine's role in putting man on the moon. The F-1 engine was used in a cluster of five on the first stage, and that was the only power during the first stage. It took the Apollo launch vehicle, which was 363 feet tall and weighed six million pounds, and threw it downrange fifty miles, threw it up to forty miles of altitude, at Mach 7. It took two and one-half minutes to do that and, in the process, burned four and one-half million pounds of propellant, a pretty sizable task. (See Slide 2, Appendix C)

My history goes back to the same year I started working at Rocketdyne. That's where the F-1 had its beginning, back early in 1957. In 1957, there was no space program. Rocketdyne was busy working overtime and extra days designing, developing, and producing rocket engines for weapons of mass destruction, not for scientific reasons. The Air Force contracted Rocketdyne to study how to make a rocket engine that had a million pounds of thrust. The highest thing going at the time had 150,000 pounds of thrust. Rocketdyne's thought was the new engine might be needed for a ballistic missile, not that it was going to go on a moon shot.

Chapter One

In July 1958, we began the International Geophysical Year, wherein the scientists of the world were going to plan and execute experiments to learn more about the earth. It was an eighteen-month program. As part of that, the United States planned to launch an artificial satellite to circle the earth, and they were going to do it within the geophysical year. President Dwight Eisenhower intervened with one portion of that. He had a fear of what he called the "military industrial complex" becoming too powerful. He wanted the space program, the satellite program, to be done without benefit of any weapons system or, in fact, weapons system personnel; so, all of the people who knew best how to make rockets go up weren't allowed to work on the project. They had to get other people, "civilians" as they referred to them. That's what happened to bring the Vanguard Program into being outside of what we had; the Jupiter, the Thor, the Atlas, the Redstone, the Navaho, all of these were already flying and almost operational. This was happening at a time of decreasing national prestige due to a number of events.

The first one was in October 1957. The world and the United States were shocked by the Soviets announcing they had put up a satellite, Sputnik I. It was a satellite weighing approximately 183 pounds. It was shocking to us that they could do it before the United States did. Before we recovered from that shock – it was within a month – they launched Sputnik II; it had a dog as a passenger. Sputnik II also gave the United States information on how well the Soviets were doing. The third stage did not separate from the orbiting package, and the third stage alone weighed 16,000 pounds. The Soviets had managed to put 16,000 pounds into orbit! Our Vanguard satellite was going to be eighteen pounds, by the way. From what was put into orbit, the United States was able to calculate that the Soviet booster must be three times as powerful as the most powerful booster we had, which was the Atlas cluster of three engines. That gave emphasis for starting to look at a much higher thrust engine. During this time period, an organization called the Advanced Research Projects Agency was formed within the Department of Defense to manage the new technology for all of the services. The Air Force, Army, Marines, and Navy were all developing the same technology. The new agency was designed to prevent that; it was going to develop the new technology for space. (See Slide 3, Appendix C)

The first attempt at launching the Vanguard satellite was December 6, 1957. It went up four feet; it blew up four feet off the launch pad. By this time, Eisenhower had relented and allowed the Army Ballistic Missile Agency (ABMA) to step in and put up a satellite. They actually used a Redstone vehicle with a Redstone engine, and they called it a Jupiter-C. Dr. Wernher Von Braun, who headed up the ABMA, let it be known that he could put one up in two months, and they finally turned him loose.[1] Very shortly after that, Explorer I was in orbit. NASA

---

[1] For more details on this, see page 111, of *Dr. Space, The Life of Wernher von Braun,* by Bob Ward, Naval Inst. Press, Md, 2005.

absorbed ABMA in March 1960 and took over all of the non-military space effort. In May 1960, the U-2 spy plane, which was supposed to be able to fly above Russia without harm coming to it, couldn't be reached. The Union of Soviet Socialist Republics had shot the plane down. The next shock came when Yuri Gagarin was put into space; a human was in space for the first time on April 12, 1961. That was followed five days later by the Bay of Pigs fiasco. John Kennedy had been president for just a few months and had allowed himself to agree to a CIA-coordinated invasion of Cuba under the assumption that all Cubans would rise up and fight against the dictatorship of Fidel Castro, and that didn't happen.

In May 1961, Alan Shepard completed a fifteen-minute suborbital flight test as the first American in space. Based on the results of that, John Kennedy made his famous speech about sending man to the moon and returning them safely. He waited until we had success in space before making the announcement. The success was this small fifteen-minute suborbital flight. On May 25, 1961, that speech was made, the same day the first test was performed on the F-1 engine. (See Slide 4, Appendix C)

The F-1 mission was to provide seven and one-half million pounds of thrust for the Saturn V first stage. It was based on, except for size, pretty much the same technology being used for the ballistic missiles. The difference was that it was a lot bigger, and the biggest challenge was just *that* – its bigness. It was ten times the thrust of the biggest current production engine, which was 150,000 lbf (pounds-force). This was going to a million and one-half. Upratings (upgrades in power, size, etc.), traditionally, even in other things besides rocket engines, are reaching if they go 10 percent. We had several 10 percent upratings before, but nothing like this. The engine was eighteen feet tall, twelve feet wide. It had a thrust level that started out at 1.5 million pounds force, plus or minus 3 percent. They made a change to it and called it an uprating by going to 1,522,000, plus or minus 1.5 percent. If you do the math, all that did was move the million up to the middle or the high part of the tolerance. The engine essentially had to be designed for the same pressure anyway. The specific impulse (a measure of engine performance efficiency) was 265 seconds, which was good for those days. It looks not so good today. (See Slide 5, Appendix C)

Combustion chamber pressure was another challenge. This was designed at 1,100 pounds per square inch (psi). The highest chamber pressure of the day was 520 psi, so this was a doubling of the chamber pressure and significantly more thrust. The engine burned liquid oxygen and Rocket Propellant-1[2] (RP-1, a highly refined form of kerosene) at a mixture ratio[3] of 2.27, and

___

[2] Rocket Propellant-1 (RP-1), highly refined form of kerosene.
[3] Mixture Ratio - Ratio of mass flowrate of oxidizer (liquid oxygen) to mass flowrate of fuel (RP-1).

it had a mission duration of 165 seconds. Well, it started out at 150 seconds, and the first two Mercury flights were 150 seconds, but they were unmanned. They had the whole stack, but they didn't have the astronauts. They made flight changes for the first manned one, and from that point on, the burn duration was 165. It was designed for a qualification life of twenty starts and 2,250 seconds with a weight of 18,000 pounds. We later ran this engine at 1.8 million pounds of thrust[4], which gave it a thrust-to-weight ratio of 100, but it was pretty close to 100 anyway, and if you stripped off the stuff that was not providing thrust, the remainder of the engine would have been a 100:1 thrust-to-weight ratio. The major features are that it had a single turbopump, and it had a shaft that was parallel to the axis of the thrust chamber. It had a liquid oxygen (LOX) pump on top of the shaft, then a fuel pump, then a turbine. It took the single turbine to run the two pumps. The nozzle had an area ratio of 16:1. It was tube-wall down to the 10:1 expansion, which is where the turbine exhaust was put into the nozzle through a skirt extension. The tubes went down to 10:1, and from then on, it was a double-wall, hot, gas-cooled nozzle extension. In the turbopumps, both the fuel pump and the liquid oxygen (LOX) pump had a double discharge, a discharge on each side of the pump. The reason for doing that was to prevent the entire head rise across the pump from becoming a delta pressure radially for a fixed load. It divided the total load by two, having two different directions.

How was the RP-1 fuel used for purposes other than just burning? These were unique things at the time. For instance, we used the fuel for lubricating turbopump bearing. Prior engines and, in fact, the original F-1 design had a separate lubrication pumping system to lubricate bearings. It also was used as hydraulic power instead of using hydraulic fluid. It was used to power the engine valves and the thrust vector control actuators, and in addition, it did not require any auxiliary starting power. (See Slides 6 and 7, Appendix C)

The F-1 turned out to be a very simple engine, but it started out with its initial design as a very complex engine. The first design, which went through a design review and was approved, had three turbopumps on it. It had a hydrazine pump, which required a hydrazine gas generator to run the turbine on the hydrazine pump. The hydrazine gas generator would run the LOX pump and turbine and the fuel pump and turbine. It had a separate pump for the LOX pump, and a separate one for the fuel, and a separate one for hydrazine. There were three pumps, each with a shaft and each with a turbine, and that was replaced during development. Actually, it never made it to the first engine test. It was replaced with a single turbopump. It also had the lubrication oil system for the bearing that was eliminated.

---

[4] F-1A was rated at 1.8 million pounds force of thrust.

There was one design that was really troublesome, and we did not get rid of it before the first test. We had to live with it for some period of time, and it was called a triple manifold thrust chamber. Picture three doughnuts stacked on top of each other; those were fuel manifolds at the top end of the thrust chamber. The fuel for the thrust chamber went into the middle manifold, then went down the tubes to cool the tubes. It went down half the tubes, then back through the other half of the tubes to another manifold. There were four trombone tubes that ran from the bottom manifold to the top manifold in order to get the fuel from that end of the injector. They had tangential inlets and outlets. It set up a racetrack in both the bottom manifold and the top manifold, and the pressure loss was horrendous, but we managed to struggle by for a while until we could get a better design. (See Slide 8, Appendix C)

On the F-1 engine schematic, the turbopump was in the upper left, with green for oxidizer and red for fuel. Since the turbopumps had two discharges, they required two fuel lines and two LOX lines going to a valve. There were two LOX valves, one on each side of the engine, and two fuel valves, one on each side of the engine as well. The hypergol cartridge in the middle of the control system held a fluid that would automatically ignite as it got into the chamber. It didn't need a pyrotechnic device to ignite it. (See Slide 9, Appendix C)

Many people have wondered how the start sequence looked. It took about five seconds for the engine to start under what is known as a tank head start[5]. It had no help from outside. It started by opening the engine control valve, which was a four-way hydraulic valve that ported hydraulic fluid to open the main LOX valve. The hydraulic system on this engine was quite robust. All of the valve actuators were way overpowered. There was no concern of hydraulic contamination with the engine. By opening the LOX valves first, the LOX started flowing through the pump in such a way that it treated the LOX impeller as a turbine, and it started to turn the pump. It would turn the pump up to 700 to 1,000 revolutions per minute, which got rid of the breakaway torque concerns and started the engine going up in power. Right after that, the gas generator valves opened. The gas generator had mechanically-linked valves, one for the fuel and one for the LOX, so they both opened at the same time. The gas generator would ignite at a very low pressure and then start building up. It would take two or three seconds for the pressure to get up to the point where the igniter fuel valve would open, allowing the fuel to go into the chamber and start the thrust chamber ignition. As soon as the thrust chamber ignition was sensed by another valve, by just measuring the fluid pressure, the main fuel valves opened, and the thrust was allowed to come on up. At about 80 percent, the thrust laid over

_____

[5] Tank head refers to the pressure at the pump inlet due to the weight of propellant in the tanks located above the engine.

for a little while and then went on up at a different ramp (i.e. rate of increase). The reason was that the thrust chamber was primed with ethylene glycol, so the initial fuel burning during this engine start transient period was an ethylene glycol and water mixture. It didn't provide as high a chamber pressure as the full RP-1 fuel so, when that ran out, the chamber pressure kept going on up and reached the 100 percent value. (See Slide 10, Appendix C)

There was a qualification test requirement to take two engines off the flight line. Supposedly, the concept was to grab two engines at random and put them through a qualification test series. One of those series was life demonstration, which demonstrated the twenty tests and the 2,250 seconds. The other one did that and also demonstrated a lot of simulated engine malfunctions and different environmental tests. These two series were quite successful and completed the contractual requirements for qualifications testing, but they were not the end of running a lot of seconds on engines. This was a supposedly expendable throwaway engine with a mission duration of 165 seconds, but we actually tested six engines to an excess of 5,000 seconds. The only reason this was a non-reusable engine was it was non-recoverable. If we could have recovered it after a flight, it would have reusable. (See Slide 11, Appendix C)

We adopted the idea of acceptance testing, using the concept of an "all up" test at every opportunity – test the whole thing. The concept was to do that with the full stage, so, when we had a series of production engines, they first would have to be acceptance tested on the engine test stand. That required two tests of duration – a forty-second calibration test and a 165-second mission duration test on every engine. After that, the engines were grouped in clusters of five and put into an S-IC stage, the actual flight stage. They then were tested for 125 seconds, which was the maximum duration they could get without having some acceleration. The first three stages were tested at NASA's Marshall Space Flight Center in Huntsville, Alabama. The rest of them were tested at NASA's John C. Stennis Space Center near Bay St. Louis, Mississippi, on the B-2 Test Stand. The total time required for the acceptance test was 495 seconds – for a mission duration of 165 seconds. All of that added up, but the highest time any engine achieved was 800 seconds. (See Slide 12, Appendix C)

There were a couple of development problems with the F-1. The first one was the most famous one. I think many people who know nothing much about the F-1 know that it had a combustion instability problem. It was thought before the first engine test – a couple of years before – that just by making it this much bigger, it would not be stable. It would have to be unstable. We ran some tests on a device called "King Kong," which was a solid-wall, steel chamber, and the tests were run for a very short duration. The test stand could only go through a start, then maybe half a second of main stage firing and then a shutdown before it ran out of propellant. Of course, this was all pressure-fed, a pretty difficult thing to arrange. The first few tests had some instability, which never was understood properly, because with the demonstration tests, everything was stable. That allowed us to make the conclusion that it probably was stable, and to go on with the plan. (See Slide 13, Appendix C)

In the first year of testing – because of other problems – we were unable to operate at rated thrust (the performance level an engine is designed to achieve). That's not unusual. In fact, it's more usual than not that the first engine in a new design is unable to achieve rated thrust for some reason. The maximum thrust the F-1 could hit for the first engine was 1 million pounds; so, for the first year, testing was limited to 1 million pounds. During that year, we got some spontaneous combustion instability strikes, but only seven times. That's a rate of about 10 percent. The tests we ran[6] would not have been able to predict whether it would have a random instability with some probability; it would only detect if it was always unstable. Well, this was a low probability, but not nearly low enough to be used. During the year, we would have instability detected by accelerometers on the thrust chamber; then, the engine would be shut down by a thing they called the rough combustion cutoff device. It was a nuisance simply because we were losing tests. It wasn't doing much damage. I guess it wasn't doing any damage because we were able to go test again.

Then, in the second year, we were able to run rated thrust for the first time. The thrust level went up to 1.5 million, and the testing continued. Then, we got a spontaneous instability that was so severe, it destroyed the engine. The initial shock was bad enough to cause both of the dual fuel lines to rip off the engine, so the engine was running with LOX and no fuel. It was burned up. It was a total loss. We had eleven such episodes in that time period. All of them did some damage. Only three of them blew the ducts off and totally destroyed the engine, but the rest of them did a considerable amount of damage. As soon as that first one hit that destroyed the engine, the problem of instability got more attention than it had previously. "Project Go" was formed, headed up by Jerry Thompson at Marshall Space Flight Center and Paul Castenholz at Rocketdyne with Bob Levine, Dan Klute, and Bob Fontaine. They made some changes to the injector. The injector had baffles sticking down, dividing it into thirteen compartments. We had fifteen different arrangements of baffles that were tested. There were also fourteen different injector configurations behind that. All of those were tested, and, eventually, the instability went away. We never again had a spontaneous instability. After a while, in the third year, we were not getting any spontaneous instabilities. However, we were getting forced instabilities

> ...in order to prove the injector was a stable design, we had to put a bomb in it, detonate the bomb, and, then, damp it out within forty-five milliseconds.

---

[6] Tests with King Kong, the solid-walled steel chamber.

because, in order to prove the injector was a stable design, we had to put a bomb[7] in it, detonate the bomb, and, then, damp it out within forty-five milliseconds. Even though the injector was not spontaneously going unstable, it was still a good requirement to make sure that if you ever did that, it would damp out immediately. In that third year, we got nine cases where we were bombing it to prove that it was stable, and it went unstable. However, a lot of the other tests were successful, and at the end of the testing, the configuration was smooth running. It then went through a complete qualification test[8] and a lot of bomb tests, and never again was there any instability. That problem was solved considerably before the first flight.

We also had another problem that was kind of interesting because it had a catastrophic effect. The LOX pump had six vanes in it. The vanes were weak. I don't want to say under-designed, but it was weak for that environment. I think the basic problem was they were scaled up from the smaller ones, and you never really had to worry much about the forces across the vane as much as this size did. It made it worse. We had failures of that vane four different times. These were four engines that were destroyed by vane failure in the LOX pump. It occurred at 110 seconds, 110.5 seconds, 107.7 seconds and, then, at 109 seconds. Statistically, this looked extremely significant. There was something going on that was a function of time, and at 110 seconds we had a problem. We did an exhaustive investigation twice and could not find any justification for that and concluded it was just a freak coincidence. We made some changes to eliminate fretting (wear and corrosion damage). What we really did was to set a limit at 3,500 seconds for the impeller for ground testing. All the flight engines never got up above 800 seconds. The last engine that went up in 1965 had 5,000 seconds on the impeller. From that, we concluded you could set the limit at 3,500 seconds and be okay. One thing about this was that on every flight, the people who were familiar with the F-1 program would notice when 110 seconds went by and, then, breathe again. (See Slide 14, Appendix C)

One significant milestone in the F-1 program was moving from the first test to the first rated duration and thrust test. It took one year to run a test that was rated thrust and full duration, and I think that's better than a lot of programs were, including some recent ones. The first flight was November 1967. I was in Firing Room 2 at the window for that flight. I considered that the highpoint of my career. I've never had such a feeling of pride, just watching the thing go up and imagining that the earth was trembling. The glass window in front of me was moving inches. It was quite an experience. (See Slide 15, Appendix C)

---

[7] A small charge to provide a pressure impulse in the combustion chamber to test for combustion stability.
[8] Qualification testing for rocket engines.

We got over 280,000 seconds of total burn time throughout the entire program. There were twelve Apollo flights that used the F-1 engines, then the Skylab used the last one that flew. That added up to thirteen flights or sixty-five total engines in flight. For those flights, those sixty-five engines were 100 percent reliable. In addition to that, we had a formal reliability demonstration program where we pre-declared the firing of the ground test engine would be one that counts and, then, went 336 total equivalent tests without a failure, all pre-declared. The engine was quite successful in that regard. (See Slide 16, Appendix C)

At times, I've been asked, "What was it like to work on such a program?" Well, it was great. It was the most important thing going on in the country, and we were in the middle of it. A lot can be said about it.

I want to share one other aspect of it. At the time we were doing ballistic missile development in 1956, the total employment at Rocketdyne was 5,000. At the peak of production of these missiles, the employment was 14,000. When Kennedy's "Man on the Moon" speech was made, there were 11,000 employees. That got a fast buildup to 20,000 in 1965. At the time of the first unmanned Saturn V flight, early in the program, employment was down to 14,500. When the first moon landing took place, we were down to 9,000, and when the last moon landing took place, we were down 2,500; 17,500 people were laid off in a few years.

The total work force on Apollo peaked at about 400,000, and I imagine the whole program went down about the same amount. (See Slide 17, Appendix C)

Editor's Note: *The following information reflects a question-and-answer session held after Biggs' presentation.*

STEVE FISHER[9]: Just one comment I wanted to make while Mr. Biggs was talking about work force population and stability in the F-1 Program. Several years ago, I had the opportunity or the task to create a briefing on F-1 stability. While I was doing that, somebody lent me an old Rocketdyne phone book. Within the Combustion Devices Group - there were several different groups: main injector, gas generator, gas generator injector, main chamber, etc. - there was a group called Main Injector Stability. All they did was stability, nothing about performance,

_____

[9] Steve Fisher served as facilitator during the *On the Shoulders of Giants* seminar series.

fabrication, whatever. Under Main Injector Stability, there were *thirty-five* names in the phone book that year.

QUESTION: Back in 1961 to 1965 when you were working with RP fuels, what was the state of the liquid hydrogen technology?

BIGGS: The J-2 (engine project) was developing hydrogen technology.

QUESTION: In 1963, the plan was changed to do "all-up" engine testing or "all-up" stage testing. Did that have any effects on the F-1 Program? Are you aware of any changes to the qualifications/certification program?

BIGGS: Well, I can say the "all-up" testing requirement probably saved us from a flight disaster because of a stage test that found a problem. The acceptance test of S-IC 11 had a large fuel leak caused by an incorrect installation. The hydraulic fluid was being supplied from an engine to the hydraulic actuators through a line that was installed on the engine by Boeing at the stage level. It went through Rocketdyne with a test plate, a good test plate cover. At that time, the line was installed for the stage test, and that would be the same as it would be for a flight, supposedly. Well, the mechanic installed a line over a cover that was left on the port. It caused a fuel leak during the test, which caught on fire. It led to other things on that test, which resulted in scrapping two engines. It would have been bad in flight. I can't say how bad because there were other things wrong on the test, but certainly, that was a condition only an "all-up" test would have found.

QUESTION: I'm Pete Rodriguez from Marshall Space Flight Center, and we're doing an assessment of what kind of testing stands and capabilities we need to have available. What kind of damage happened to the actual test stands during the instability failures, and how long did it take to fix and to get back up and running?

BIGGS: Engine test stands, in general, are quite robust because if they fail and come apart, that energy is lost in destroying the engine. The test stand can be damaged by fire, but I think most of the damage cases were wiring and similar items. In some cases, a lot of damage can be done. But in the instability of the F-1, the worst it did was to blow off the propellant supply lines. That was pretty bad, but it kept the additional damage local to the engine.

FISHER: I think typically, there is a lot less damage in an engine stand than in a high-pressure stand, as you can imagine. Obviously a lot of repairs would need to be done, but most infrastructure remains.

One of the J-2 engines that power the upper stages of Saturn V liquid hydrogen, on its way from the fuel turbopump, is used to cool the walls of the thrust chamber regeneratively. (Source: http://history.nasa.gov/SP-350 - Apollo Expeditions to the Moon", published 1975)

Chapter Two

Paul Coffman
Rocketdyne - J-2 Saturn V 2nd & 3rd Stage Engine

Paul Coffman earned experience during many Apollo-era assignments. He served as lead engineer on the J-2 thrust chamber assembly development, supervisor of engineering test for J-2 components and engines, and manager of J-2 engine development and flight support. Following the Apollo era, some of Coffman's program management and business development assignments included serving as director of controls and engineering, program manager of the compact steam generator, engineer coordination for the space shuttle main engine, and project engineer on the gas dynamic laser. He also led successful proposal efforts on an advanced storable propellant spacecraft engine and a system test bed for potential space station propulsion.

The J-2 engine was unique in many respects. Technology was not nearly as well-developed in oxygen/hydrogen engines at the start of the J-2 project. As a result, it experienced a number of "teething" problems. It was used in two stages on the Saturn V vehicle in the Apollo Program, as well as on the later Skylab and Apollo/Soyuz programs. In the Apollo Program, it was used on the S-II stage, which was the second stage of the Saturn V vehicle. There were five J-2 engines at the back end of the S-II Stage. In the S-IV-B stage, it was a single engine, but that single engine had to restart. The Apollo mission called for the entire vehicle to reach orbital velocity in low Earth orbit after the first firing of the Saturn-IV-B stage and, subsequently, to fire a second time to go on to the moon. The engine had to be man-rated (worthy of transporting humans). It had to have a high thrust rate and performance associated with oxygen/hydrogen engines, although there were some compromises there. It had to gimbal for thrust vector control. It was an open-cycle gas generator engine delivering up to 230,000 pounds of thrust.

We delivered 152 production engines. The specific impulse of 425 seconds did represent a small compromise. Chamber pressure was relatively modest, a little over 700 pounds per square inch absolute (psia), and the mixture ratio was 5.5:1 at that chamber pressure. It also had the capability of operating at a mixture ratio 4.5:1. Basic engine weight was relatively light, and if you make the numbers, you will come out a little bit over 80 percent for thrust-to-weight ratio. Propellants were liquid oxygen and liquid hydrogen. Nozzle area ratio was 27.5:1. Because of the difference in densities, there was a requirement for two different turbopumps. They were separate and located, as conventional wisdom would have it, on opposite sides of the engines just for safety sake. The bearing lubrications were liquid oxygen and liquid hydrogen, so there was no separate lubrication system. The gas generator burned main propellants. (See Slides 2 and 3, Appendix D)

As far as the engine operating in the main stage, there was a gas generator that was fed with fuel and oxygen off the main propellants ducts. The gas generator drove the turbomachinery, in a series turbine arrangement, through the fuel turbopump and then over into the oxidizer turbopump, with a bypass for calibration and a heat exchanger to heat up oxygen for tank pressurization (or helium in some instances), before dumping into the thrust chamber at the nozzle midpoint. It was a fully tubular thrust chamber with a fuel inlet at the reduced epsilon[1] portion of the nozzle. It was one tube down for every two tubes up. There was a 2:1 split, and we used the opening at the 2:1 split to dump the hot gas into the nozzle of the thrust chamber. The oxidizer turbopump was a fairly conventional centrifugal device. There was a rather strange shape for the fuel turbopump. The fuel turbopump was an axial machine. It was derived from machines that had been used in the nuclear rocket programs. While that was deemed a sound idea at the start, it resulted in probably the single major development problem – a start sensitivity for the engine. (See Slide 4, Appendix D)

Again, the engine had to restart. It started at altitude on the S-II and S-IV-B first burn and, then, on the S-IV-B restart after one and one-half to two and one-half orbits. The energy for start was supplied by cold hydrogen that had been loaded either on the ground or reloaded from the main fuel injection manifold during the initial burn operation, and it was this start tank that would discharge cold hydrogen through the two turbines to get the engine started. Inside that start tank was a small helium tank because the engine carried its entire helium supply with it in a cold helium tank.

---

[1] Epsilon refers to nozzle area ratio, exit area to throat area.

Altogether, 152 engines were produced. The engine development, while it may have ceased with the qualification efforts, continued with the F-1 and persisted throughout the entire program. It also spawned a J-2S engine and the Linear Aerospike Engine test beds that were operated at Rocketdyne as well. From the start in 1960, the first 250-second duration test occurred a couple of years later, not a year as the F-1 managed. The 500-second full duration test happened another year after that. We then began a series of tests to demonstrate formally the engine readiness through preliminary flight rating tests (PFRTs). Those were followed by flight readiness tests (FRTs), followed by qualification tests I (qual). Those tests were on the 225,000-pound version of the engine, fifty-nine of which were delivered. Qual II tests came in 1966. Those overlaid with the first production engine delivery, the first flight of the S-I-B, and the first Saturn V flight in 1967, with the moon landing coming in 1969. At the time we landed on the moon, we were very close to delivering the last of the J-2 engines. Perhaps interestingly enough, from a contract standpoint, the engine deliveries were incentivized in the initial contract, so there was a bounty on delivering them on time, even though the capsule-related delays that were experienced by the Apollo Program caused them to be delivered to warehouses in many cases. (See Slides 5 and 6, Appendix D)

We had thirty-eight development engines. This was not a hardware-poor program. This was a schedule-driven program; cost was no object. We had 1,700 tests through qualification. That was the qual II series. We continued testing and had over 3,000 tests by the time the program ended. We had multiple test facilities and a lot of component facilities as well. We had five engine test stands at Rocketdyne Santa Susana in California, including one that was a simulated altitude facility. We had test facilities at NASA's Marshall Space Flight Center in Alabama. We had test facilities at NASA's Stennis Space Center in Mississippi, and one at Sacramento, California, for the S-IV-B. We also did some very significant testing at the J-4 cell at the Arnold Engineering Development Center (AEDC) in Tennessee. That was significant because that was where we finally resolved the start problems that were the major issue with the engine. (See Slide 7, Appendix D)

The engine was required to complete thirty tests and 3,750 seconds of operation for the formal demonstrations. If you added up all of these numbers, it came to something over forty tests because there were more than one at a time. Indeed, we managed to do that to the point that some of them were just start-stop tests as we actually finished the program. The most significant ones were what we blandly called "the safety limits tests." Those were the extremes of the boxes for start energy and build up energy. The actual demonstration engine was Engine 2072, which did all but the last altitude simulation test. The program office was very, very conscious of, and was not comfortable with, the fact that we ran 57.4 seconds over the requirement.

There were four key development issues. (See Slide 8, Appendix D) The first was thrust chamber ignition detection. It was a good idea to have the thrust chamber lit before the main propellants were added. The thrust chamber was ignited by an augmented spark igniter in the middle with two spark plugs that used fuel and hydrogen and oxygen. It also had a spot for an ignition detection device. The development of ignition detection devices was quite significant in that the simplest device was a fusible link. Burning the link showed there was a fire. This process was used for most ground tests, except at AEDC, an Air Force-owned test site, where we wanted to run multiple tests within a given air-on period, or vacuum period. At this point, it seemed advisable to have an ignition detection device for all tests subsequent to the first one. There was also an idea that we would like to detect ignition on flight engines as well, since a reusable probe underwent a lot of development. That was used for the AEDC engine testing, but it never demonstrated enough reliability for flight. The net result was that I don't believe there was ever a ground test conducted without an ignition detection device, but none of the engines were flown with one. *That's a point to consider as we move forward to a new crew launch vehicle*[2]. (See Slide 9, Appendix D)

> The engine start was a significant development issue, and one we didn't do tremendously well.

The engine start was a significant development issue, and one we didn't do tremendously well. I think Dr. Richard Gilbrech[3] mentioned that, in this era, the analytical capabilities and modeling were somewhat limited. That was probably a very positive statement, but a sensitive one as well. There was a very strong potential for gas generator burnout if you didn't tiptoe through the start transient very carefully. There were several components that contributed heavily to that sensitivity. The first was the axial fuel turbopump. Fuel turbopumps of that configuration have a tendency to stall if there's too much downstream resistance. That required a great deal of thrust chamber assembly thermal conditioning on the ground beforehand, and a variation in fuel lead time, depending on how long it had been since it had been conditioned on the ground, in order to make sure that it was not too cold or too warm.

A complex two-position main oxidizer valve was utilized. It featured a pneumatically actuated butterfly valve. To operate, the valve had to be moved from the closed position to a fourteen-degree position and the flow had to start to begin the engine. It had to dwell there for about one-half second, then go to the full-open position required during one and one-half seconds. It is characteristic of butterfly valves, however, that if the flow force increased too rapidly on them, they would just stop moving, unless the actuator force was sufficient to move them. Now, if the actuator force was pneumatic, a force balance issue occurred.

---

[2] Italicized text represents "lessons learned" by the conference presenters.
[3] Dr. Richard Gilbrech served as director of Stennis Space Center from January 2006 through August 2007. He recently returned to Stennis Space Center as associate director.

The third issue was the turbopump start energy from the high-pressure cold hydrogen. The dump of that energy had to be precisely accommodated, and it had to be both the requisite temperature and pressure in order to get the satisfactory start energy. That process was controlled by the start tank vent relief valve. People have wondered why we used two valves. Well, we knew there was going to be a warm-up of the cold hydrogen during the boost phase, and we wanted to compensate for that. We had a big valve that was going to dump immense quantities of cold hydrogen, and we had a leak rate that had to be calibrated as well. With all of these issues combined, extensive thermal conditioning was needed. We were conditioning a thrust chamber with cold helium on the launch pad. (See Slide 10, Appendix D)

The initial problems were encountered at Santa Susana and at Marshall. The simulated altitude testing at the Santa Susana Vertical Test Stand 3, 3A position, was a small capsule. The engine was mounted horizontally, and it had a steam ejector to pull the initial altitude. It really didn't do a very good job of simulating the start transient with the spit-back. When we started flying the S-I-Bs, *which shows the wisdom of flying early if you possibly can*, we indicated the problem was not very well understood, and certainly was not resolved. Bottom line: we started putting immense activity into AEDC's J-4 cell. For those who are not familiar with it, there was a hole in the ground about 100 feet in diameter and 300 feet deep. We could attach a very strong vacuum to it. There was a huge bell jar configuration metal container, inside of which very easily fit an S-IV-B stage with a J-2 attached. We did a year and one-half worth of testing there and, concurrent with that, fruitfully advanced our analytical modeling capability, both at NASA and at Rocketdyne. The net result was that we were able to achieve satisfactory test starts on all the ground and flight applications.

The engine started relatively slowly. It could go from 10 percent to 80 percent in about one and one-half seconds. Although it was relatively slow, it was quite repeatable. The bandwidth was fairly repeatable for 152 engines, which was pretty tight, especially considering that the inlet conditions to the engine depended to some extent on the stage application. Again, we were able to replicate this and advance the cause by advancing our analytical modeling. (See Slides 11 and 12, Appendix D)

Another interesting little problem we encountered was high transient sideloads. We realized that to get at a twenty-seven and one-half epsilon nozzle and still facilitate testing at sea level conditions, it was going to take some doing. That was a goal. We used a conventional nozzle, and at a 27.5:1 ratio, it was below the three-tenth level. We assumed we were going to have full flow in the nozzle. We wanted to accomplish that, so we developed an adverse pressure gradient, or APG, nozzle that would be back up over that range as a result of over-expansion in the nozzle. There was about a two- to three-second specific impulse performance loss attributed to that. We proved the concept with both analysis and model hot fire testing. We had tenth-scale, solid-wall nozzles of several configurations that were evaluated before the APG configuration was chosen. However, once we started testing, we noticed some fairly high sideloads during start transients. Similar sideloads could occur also when thrust was reduced by changing the propellant utilization to run at the 4:5 mixture ratio and dropping the thrust level about 20 percent. The thrust chamber damage that we incurred was rather dramatic. The attach points for the actuators on the J-2 were at the upper end of the engine, and the clevis pins started elongating the holes. That may have been due to the fact that we were putting more than 100,000 pounds of pressure through them, and they weren't designed for that. (See Slide 13, Appendix D)

The second thing was spotted by an associate of mine in combustion devices at the time. He stopped to tie his shoelaces after he looked at the thrust chamber to make sure the injector was okay. When he happened to glance back over his shoulder, he noticed the exit of the thrust chamber had a unique configuration. It was no longer straight; it was concave, and he said to himself – being a MIT grad – "There's something wrong there." There's just no discounting a good education. The net result was a panic because the conventional diffuser to attach to the chamber was going to be approximately twenty feet long, and that wasn't going to fit real well into the test stands of the day. The original configuration had a very clever turnout at the tail end of the tubes. We began with the subscale model that had been used to define the contour in the beginning, and we started testing, first with wood, then eventually with metal. We finally settled on something that was about a quarter of an inch long on the subscale model – six inches long on the actual full-scale nozzle. That eventually evolved into a water-cooled version. We reinforced the hat-bands by cutting the top out of them, and welding in tubing. We also attached clevis pins to the area where the fuel injection and hot gas manifolds, which allowed us to tie into a sideload attachment mechanism, or SLAM, system. (See Slide 14, Appendix D)

There was a strange device hanging on the backend of the chamber. We put it on to allow us to get through start transient. We "belt-and–suspendered" it for both single and multi-engine ground tests, which solved the problem with no further difficulty. We needed to have the capability for this to release during main stage, so we could demonstrate gimbal. If we didn't have to demonstrate gimbal, we would usually just leave it attached at all times. As an aside, the initial version of this cheap metal monstrosity was welded to a combustion devices chamber on a test stand in the area, a pressure-fed test stand. The first test had been quite satisfactory. We planned on coming back the next day and testing it more. However, when we got back, there was nothing but a jagged edge. *The engine systems people had been over and cut it off and it now was welded on the engine next door.*

Now, the final major issue of the J-2 came when the Apollo 502 vehicle was launched in 1968. It was the second launch of the Saturn V vehicle. It was unmanned and was planned to be the last unmanned launch. One of the five J-2 engines cut off prematurely, and the S-IV-B engine operated apparently satisfactorily on the first burn, then failed to restart. We really went after that one. Flight data indicated there was probably a similar failure mode that caused both malfunctions. Hard as it was to believe, there were certain similarities in the engine area temperatures and in the engine operating characteristics that led us to believe that it was very likely the case. What we wanted to do was verify the anomalies with ground tests, define the failure mechanism analytically at the same time, verify it, and *then please, please, please incorporate corrective action, as soon as possible.* (See Slides 15 and 16, Appendix D)

A good example of this was the S-IV-B stage timeline. The engine experienced five engine tests and two stage acceptance tests prior to flight, and it had been pretty nominal. We saw engine chilling beginning about sixty-five seconds into flight. Then, there was a heating that began about forty seconds after that, and we got into chilling again during the restart. We saw very small performance shifts of about 4 psi and 12 psi. We also saw a little bit of change in the required thrust vector control. There was a slight shift in actuators that indicated something changed in the nozzle. Everything else was dead nominal. As a matter of fact, we noticed that we turned loose the start tank discharge valve and waited a number of seconds before there was a second burn cutoff initiative because it didn't go.

We started looking at the augmented spark igniter. The fuel line to the augmented spark igniter had a single-ply bellows, and we said, "Feed until that bellows fails." The idea was if that bellows failed, it would account for burnout of the augmented spark igniter, and that would certainly mirror, or have the possibility of mirroring, the chilling and heating that we observed. We set up an engine at Santa Susana to investigate that. We planned a sequence of events where we would have sixty-five seconds of normal operation, then we would incur a small augmented spark igniter fuel leak. We would increase the leak subsequently to complete the failure and allow backflow of the augmented spark igniter combustion products for the final thirty seconds of the operation. I had never been so happy to see copper in a flame of a rocket engine in my whole life. It went right down the line. There was substantial erosion of the device. The cavity burned out. The performance loss was similar and so was the thrust vector change. We figured we had it nailed. Within a day at Marshall Space Flight Center, the S-2 engine that had been set up and tested showed the same characteristics, so we figured we had done a good job in duplicating the failure on the ground. The analysis did indicate that the single-ply bellows could fail and that probably liquid air, since there was hydrogen passing through it on the ground, could damp the oscillations. That hypothesis was tested on similar

> The engine and the analytical tools utilized to develop it enabled us to be satisfactory on the space shuttle main engine, the RS-68 engine, the X-33 engine, and what is now being characterized as the J-2X engine.

bellows both with inert fluids and liquid hydrogen, and showed the failure. Some people have asked why the bellows were there in the first place. Well, our analysis initially indicated the bellows was a wise idea because of thermal changes that were going to occur during engine operation. So, the bellows were put in to compensate for that. In hindsight, a good idea turned out to be a bad idea. We got rid of the design modification. Solid-line Class 1 welds, and the revised configuration was qualified and incorporated on all the engines in the field. Eight months later, the first manned Apollo 503 flew – a pretty good response to the issue. (See Slides 17 and 18, Appendix D)

We began this project in 1960, and it was schedule-driven. It was what we would characterize today as a low technology readiness level. The thermodynamic characteristics of para-hydrogen were under investigation by the National Bureau of Standards[4]. Given the criteria, the development program was quite satisfactory: two years to the first 250-second test, formal development completed in six years. The J-2S and linear spawned off from that, to investigate and keep people sharp. There were over eighty modes of unplanned cutoffs. The engine and the analytical tools utilized to develop it enabled us to be satisfactory on the space shuttle main engine, the RS-68 engine, the X-33 engine, and what is now being characterized as the J-2X engine. (See Slide 19, Appendix D)

---

[4] Currently known as National Institute of Standards and Technology.

Editor's Note: *The following information reflects a question-and-answer session held after Coffman's presentation.*

STEVE FISHER: I'd like to introduce Mr. Manfred "Fred" Peinemann, a guest from The Aerospace Corp. Please stand up, Fred, so we can pinpoint you while you ask your question.

QUESTION: Did you encounter interesting materials compatibility problems during the development with the J-2? Did you encounter hydrogen embrittlement?

COFFMAN: We didn't encounter hydrogen embrittlement. The first problem we encountered was the first summer we started trying to do engine testing. We found that we couldn't get liquid hydrogen through the engine because the facility dump system and the engine bleed valves weren't big enough to allow it in a Santa Susana summer[5] to achieve liquid at the inlet to the engine.

QUESTION: This goes basically to both F-1 and J-2. Can you comment on the philosophy when you had failures and anomalies? How fast you turned those around? From what we can tell, it looks as though you were continually testing. You didn't stop and have failure investigation teams, and factor all that back in. How did your process for getting that corrective action work into the design occur?

COFFMAN: It is true that we were constantly testing, but there was a formal Unexplained Condition Report (UCR) system. When a component failed in an engine test, or in a component test, the UCR was generated. The first thing we had to determine was whether the latest configuration had failed, or whether it was something else that had already been replaced on the engine and the failure wasn't useful. There was enough hardware richness that we had five engine test stands, running two shifts. I remember one of the technicians looked up and said, "Wow, Friday. Only two more work days until Monday," which characterized the era pretty well. The attempt was to incorporate any corrective actions immediately, and put out kits for retrofit of everything in the field. The logbooks of which kits applied to which engines were pretty interesting. But, certainly, we had a very strong cadre of developmental personnel, at both the component and engine system levels, and no problem was ignored.

---

[5] Meaning the hot climate in north Los Angeles desert.

BOB BIGGS: I think for major problems we have always - we did in the F-1 program, we do in space shuttle main engine program - established an investigation team with an autonomous team leader to run the investigation. We assigned on the order of twenty to thirty people to work the problem, and it is generally totally resolved within thirty days. Today, we would not test for a period of time. I think on F-1 that was rare. We would continue testing while the investigation was going on.

QUESTION: Paul, from your talk, it appeared that you did a lot of testing over at Santa Susana and, obviously, we all know that the facility is not in great shape right now. You mentioned you had as many as five test stands going at the same time. Do you have any recommendations for as to how to get prepared for those tests?

COFFMAN: I think you are starting from a much more advanced position with respect to looking at developing future engine and stages. You also are not pushing the envelope nearly as much. So, I think that from what I have seen of the test planning for the J-2X, for example, I think that the group is well-connected between Marshall Space Flight Center and Rocketdyne in some of the net meetings where I have been an observer to make sure they do a thorough job of testing. Now, the question of "thorough" will always have different meanings to different people. *I would encourage, until we are very satisfied with the start characteristics of that engine at altitude, that we keep open the vacuum capabilities.* I think the test planning is headed in that direction.

FISHER: This is my opportunity as a moderator to throw in my two cents. I think, as Paul mentioned, we have come a long way. Our analytical tools, especially our transient modeling and understanding of the processes, are much better than they were. However, I think that with some of our recent successes in engine programs, we have done very well. We tend to get a little bit optimistic on a no-failure plan. I think one of our weaknesses over the years is we haven't planned for failure. If you are in a program, and you have a lot of extra hardware, and you have money to resolve it like we did back in those days, that is one thing. In today's world, where we plan for every engine test being a success, that may rise up to bite us. *So, for my two cents, a little bit of reserve is required for those failures that are unknown. It would be nice if we could carry that. I don't know how you go get that.*

QUESTION: I'm Thomas Carroll. I work systems for J-2X testing. Were there any compromises that were made for any of the previous J-2 programs, other than the planning for failures? Are there any compromises that were made that we are looking at fixing for J-2X that we want to fix, that we want to change, and want to make better?

COFFMAN: I don't think so, at least as the engine is evolving now. There seems to be a rather thorough study of the "S" (i.e. J-2S) and the basic J-2. You must recognize that the "S" was commissioned to be physically and functionally interchangeable, so it had to meet the 80-inch diameter, and the same length, and the same attach points, and it had to fit the same buildup characteristics, and so forth. There may be some real possibilities for engine definition for the "X" (i.e. J-2X) in coordination with the stage to be mutually advantageous, but I don't have any concrete examples of compromises that would cause a problem going forward.

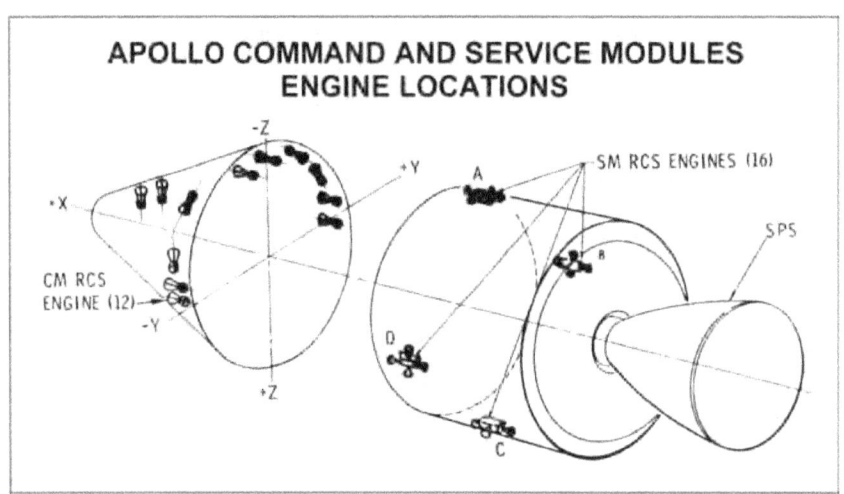

Apollo Spacecraft/LM Adapter (Source: Apollo Training Manual "Apollo Spacecraft & Systems Familiarization" – March 13, 1968)

Apollo Reaction Control System Engines (Source: Pfeifer presentation charts, Appendix E)

Chapter Three

Gerald R. Pfeifer
Aerojet - Attitude Control Engines

Gerald R. "Jerry" Pfeifer joined the Marquardt Company[1] in 1961, performing early work to develop the parametric rules used for Marquardt rocket design and teaming with a small group to build the company's first hypergolic rocket test facilities and rockets. When the Apollo reaction control system project came to Marquardt, Pfeifer helped shepherd it through its qualification phase. The work led to involvement with many other successful space propulsion programs, including work for NASA's space shuttle. Pfeifer eventually went to work at Rocketdyne to help develop its Terminal High Altitude Area Defense (THAAD) attitude control engines. He now is working on the Airborne Laser Program at The Boeing Company.

In late 1960, I went to work for Marquardt, which was an interesting company. *It was a big, little, big, little, big, little, gone company*[2]. When I went to work there in 1959, they had 5,000 people. I went to work on a program called Bomark, which was an interceptor missile. That program was downsized, and we had a chief engineer named Warren Boardman, who had this great idea after Sputnik that there had to be a future in space somewhere, and since we were a propulsion company, we would probably be good at something like that. We went off to figure out what it might be. He convinced the company to give us some internal research and development, and they turned five of us loose to figure out what propellants were and to try and figure out what the parameters were that could be used to make one of these silly things. (See Slide 2, Appendix E)

Boardman had a really novel idea that instead of what the industry was headed for at the time –

---

[1] Marquardt Company of Van Nuys, Calif., is no longer in existence. Carl Stechman is also mentioned in this article as a collaborator of the speaker at Marquardt.
[2] Italicized text represents certain "lessons learned" emphasized by the conference presenters.

many different sizes of engines where you would run long durations to control attitude in space – if you pulse-modulated the things, you could do a whole job with thrust levels from five to about 100 pounds. We developed enough credibility in that early time that we managed to win a little program called SINCOM. In that program, we had two thrusters, with a single unlike doublet injector bolted assembly on both of them. Both the twenty-five-pound and five-pound single unlike impinging injectors[3] were what we had been working with in the backyard. The backyard was really what it was. This little group that we put together built our own test facility, and we made one mistake. Basically, the facility consisted of a six-foot high concrete wall with propellant tanks on one side, and two guys on the other side looking through the window to see what happened. The mistake was we didn't remember that the wind blew from the west, and we were on the east side of the test stand. The engines weren't very efficient, so, frequently when we evacuated, we went through this red cloud.

However, that program gave us enough credibility that we were able to convince North American Aviation Inc. and NASA to give us the contract to build what they called a common engine. It had three attitude control applications: the Apollo service module, lunar module, and the Apollo command module, which was to be a varied engine. We got about a year into the program, almost a year, before everyone came to the conclusion that this little company, which had gone from 5,000 to 500 employees really couldn't handle a large program of that magnitude; so, the command module part of that went to the Rocketdyne Company. (See Slide 3, Appendix E)

My colleague Carl Stechan started at Rocketdyne and came to Marquardt. I started at Marquardt and ended up at Rocketdyne. We did all of our development work with single doublets,[4] and one of us was convinced that a 100-pound single doublet could work at the 100-pound level very satisfactorily. It really wasn't too bad an idea, except the chief engineer said we had to have a specific impulse of 300; 270 didn't cut the mustard. The thing evolved then into what it is now and started out from the original configuration, the R-4A. The R-4A was an eight-doublet with eight film cooling holes and two valves. The fuel valve canted off to the side to a circumferential manifold, and the oxidizer valve fed right through the center of the injector to eight outward pointed doublets. It featured a molybdenum nozzle, coated with molybdenum disilicide or Durac-B. (See Slide 4, Appendix E)

Basically, we had three groups working. We had a guy doing the injector and combustor design.

---

[3] Injector with a single element (i.e. unit) consisting of one fuel orifice and one oxidizer orifice that impinge the resulting fuel and oxidizer streams together just downstream of the injector face.
[4] A type of injector where two fluid jets impinge to create a spray.

Carl Stechman was doing heat transfer work. We had another group doing materials studies. In actuality, the molybdenum system they came up with wasn't too bad except for two features. One feature was the material, which, if not coated, vaporized when it got hot. It was very, very sensitive to chipping. The way we tested the nozzles to find out if they were okay was to put them in the oven at 2,500 degrees; if they smoked, they were gone because you could see them while they smoked. They were also very brittle. It's a very brittle material. The ductile brittle temperature was about 60 degrees Fahrenheit, so once they got cold, they were like just tapping a piece of glass and, with our shocks, that was not a good thing. In order to keep an engine in one piece – because the nozzles kept splitting – we got the R-4B configuration in which we put on a ribbed L-605 nozzle attached to the molybdenum combustion chamber. That continued to grow, and they all looked the same. From the outside, they were all pretty much the same. (See Slide 5, Appendix E)

The original eight-on-eight injector basically stayed the same from the first injector we put together in 1962. The valves improved. Scads of things changed. However, the Preliminary Flight Rating Test (PFRT) injector differed only from the qualification injector in flight units in the valve seat materials. We were having difficulty with the high temperature soak back after long runs. We decided it would be a good thing to strengthen the Teflon valve seat. In order to do that, we put fiberglass into the Teflon valve seat, and it really stayed together, except when the engine heat soaked back; the Teflon grew and grew and grew, and the oxidizer didn't get through the oxidizer side, if it got too hot. We flew two successful missions around the moon with that PFRT engine. Boeing made that particular device. We held our breath all the way because it was used as the brake engine and an apogee engine both; we were really worried about the valve seats getting too big. In the middle of the chamber of the basic R-4D engine as it evolved into the engine that flew on all the missions, there was a little tube sticking down, and that was a magic thing called the pre-igniter. It was both part of our problem, and a savior of part of our problems. The thing was a cooled molybdenum chamber with an L-605 nozzle. The reason we stuck the L-605 on was that it was a high temperature material. The ribs had to be on there to keep it round. It also had one other nice feature on the service module, they stuck out in all directions. The flight engines all ended up with a no-step design on the nozzles because they turned out to be convenient. (See Slides 3-5, Appendix E)

> … a magic thing called the pre-igniter. It was both part of our problem, and a savior of part of our problems.

On the lunar orbiter, the 100-pound engines faced down. They used cold gas for attitude control. That was a pre-qualification test engine. We did have some technical difficulties. The first one was high heat transfer burning. Today, I can confess it was first-tangential combustion instability mode, but at the time this happened, it was really an interesting thing because in the Apollo mission duty cycle,[5] it had to perform a ten-second run followed by a very specific off time, then a three-second run followed by another thirty seconds, and then ten seconds of continuous burn. The characteristic was they would go through the first two very nicely, and on the third burn, in about one and one-quarter seconds, the temperature on the chamber wall would pass 3,100 degrees Fahrenheit. Shortly before that happened, the fire would come in a thrust vector mode, which wasn't planned. Actually, we compromised with NASA on that one. We just decided that was probably not a good duty cycle because nobody quite knew what it was or how to cure it. We compromised and went through a different duty cycle to get around that particular problem, which was the only place that it really occurred on that engine. (See Slide 6, Appendix E)

> It took years for us to figure out what a good thing we had done without planning it.

We had a series of difficulties with spikes. That engine had some neat things to get through. It was generally operated in ten milliseconds on-time duty cycle. Thermal control was something we thought about, but the combustor in that mission had to sit there and, characteristically, it would go down to minus-100 degrees Fahrenheit. When the engine shut off in short pulse duty cycle in that kind of a radiative environment, the oxidizer entered the manifold first because it had gotten a very high vapor pressure relative to the fuel, and it went to the closest cold spot, which was at the chamber wall, and the fuel would come dribbling out. It also looked for a place to stop. It would freeze right on top of the oxidizer. We would build multiple layers of fuel oxidizer, and then, we would get a run long enough up to really do something. We'd get a really exciting event. I deemed it "pre-unplanned disassembly" at the time. That problem really went away when we learned a little bit about fuel control and kept injectors and propellant temperatures a little warmer. Aerozine-50 (A-50) propellant was the original propellant we selected because the big engines that had to be used to get us to the moon all used A-50, so that had to be the propellant of choice for us. (See Slides 7 and 8, Appendix E)

Carl Stechman didn't think the pre-igniter did something, but someone else did, and the NASA folks kind of agreed. The pre-igniter really was a good thing, but the serendipitous thing it did was the hot phase or first tangential instability went away because we had changed the interior configuration of the chamber a little bit. It took years for us to figure out what a good thing we had done without planning it.

The other major technical issue was the inner manifold explosions resulting from monomethyl hydrazine (MMH) evaporating and depositing like dew in the cool spot in the engine. That

---

[5] A parameter referring to time-varying, on-off cycling of any rocket thruster or engine (also to intermediate states of thrust, if needed).

was a recurrent problem, and it was the reason the space shuttle has a limit of 70,000 feet operation today. In space, it can't happen because there is no way to get the phenomena to occur because the pressure is inside the chamber and injector after engine firing, and it is less than the thermodynamic triple point of the fuel. We did a lot of premature disassembly (i.e. failures). The chamber would disintegrate during pulse mode firings in ten-millisecond pulses. The thing that really convinced us to drive to the pre-igniter was the program manager's data plot. After a night of testing, he had drawn a plot that showed chamber wall thickness across the bottom. The ordinate was the peak value of the measured pressure. He had infinity drawn both directions. It seemed like that might be a wise thing to change. Later, the pre-igniter really did alleviate the spike problem to a high degree. When we changed to warmer operating conditions, like a 70-degree Fahrenheit propellant, the problem went away and never came back.

In later developments, we put in a columbium chamber, which was a lot more tolerant. A-50 going away was also a good deal because it was more sensitive to that chamber. There was a lot of resistance going to the niobium chamber (niobium and columbium are the same material) because we had a molybdenum chamber that worked. If it was good enough and if it ain't broke, don't fix it. Stechman wrote an article about this feature in the *Journal of Spacecraft and Rockets*. It was not a real detailed discussion, but there is some discussion there. (See Slide 8, Appendix E)

Another problem we ran into was a thing called a ZOT. There used to be cartoon called *BC*. A couple of the key characters were an anteater, who was always getting after ants in the ant hill, and a snake, who always got struck by lightning when he would do something inane. The ZOT was something like that snake. We didn't really understand it too well, but it sure was something. It wasn't called the ZOT originally. It was called, "What the hell caused that?" (See Slides 9 and 10, Appendix E)

If you could picture the oxidizer valve off in one of the early Apollo engines after it disassembled itself from the engine, there was a little squirrel thing running around in there; that was the seal between the injector and the valve. We thought something happened; after it happened a couple of times, we began calling it a ZOT, because, like that cartoon snake's observation, something was really interesting. Its characteristics were pretty definable. Ordinarily, when you opened a valve, you got a smooth pressure decay. Instead, what would happen was the pressure at the instant of the valve opening would disappear off the chart. We finally associated that with high pressure. That was the clue that we were looking for, and we finally figured it out after somebody asked us to run some vertical, facing-up engine tests. We discovered that it really was kind of an interesting phenomena confined to a relatively small operation area. We

Chapter Three

made a ZOT plot written by Stanley Mitten, who was one of the better engineers that I dealt with in the science part of the world over the years. Basically, it said that at very low pressure, there were no ZOTs because the fuel emptied the manifold and evaporated really quickly and was gone from the chamber almost instantly. (See Slide 10, Appendix E)

At very low pressures, it wasn't such a good deal – particularly in the pulse duty cycle, which was pretty hardware dependent. If there was a place in the engine between the valve seat and the injector face, that could be cold. Guess how it got cold. When the oxidizer evacuated the injector, all these small engines had the valves separated from the injector face by a thermal standoff. If the valves seats could be kept cool, while the injector face was running very hot, the little thin standoffs in between the valve seat and the injector could get very cold when the oxidizer was evaporating and emptying the manifold. The fuel came out after, and if the ambient pressure was just right and the ambient pressure was above the triple point pressure, dew would develop inside the oxidizer manifold. If enough dew collected, on the next pulse, liquid fuel would develop inside the oxidizer manifold, and it would generate some horrendous pressures. We'd get premature disassemblies[6] again. It wasn't a problem for Apollo because that particular engine never ran in real life in its missions until it was well out in space. This wasn't a space problem; it was purely a ground problem. It was a problem for the space shuttle because the shuttle engines operated for attitude control during the early phases of reentry. There was a big concern because through stage down to 70,000 feet, there is the risk of having a ZOT occur and a valve leak or too much condensed fuel in the oxidizer manifold harming the hardware pretty severely. The ZOT story, today, is probably still a problem that people need to think about for the upcoming vehicles, because the shuttle engine is going to be, with any probability, one of the attitude control engines for the new large vehicles. The 870-pound thruster is the ideal size for intermediate control. This is one of those problems now that we recognize, we know what it is, and we know how to deal with it. In the current shuttle engine, that is an easy thing to fix. (See Slides 11 and 12, Appendix E)

People often wonder about schedules. Marquardt had no schedule problems at all. Our task was really pretty easy. We just had to do a little development, which happened in the 1962-1963 time frame, and then, we did a little more development. Then, we got the R-4D, and we did some more development. By the time we got out to the 1965 time frame – and that is a September 1965 schedule as a matter of fact – we had managed to make a little bit of hardware and break quite a bit of it. We were just getting into the qualification program, and we had this small task to make 700 engines. Interestingly enough, once we decided we had the problems

> We were just getting into the qualification program, and we had this small task to make 700 engines.

---

[6] Meaning that the thruster exploded.

behind us and were pretty well along the path of making hardware that we believed was okay to support manned flight, we geared up and produced about ten engines a week out of that little factory, which had grown from 500 to about 800 people during the Apollo Program. There wasn't a straight-forward acceptance test that consisted of four pulse-firing test series. There were two ten-second runs to get good performance data. Then, we had to run sixty-second durations, so we could demonstrate engine life margin. On average, we ran two of those engines a day for two years to get all of that hardware out of there. It was an interesting time. (See Slide 13, Appendix E)

The cost was always an issue. The hardware evolved up to be an Aerojet product today. That same R-4D heritage resides up at Aerojet Redmond now. Carl Stechman is their corporate knowledge of what the history of that engine is and really knowledgeable about what makes those engines work. In 1960, the price of an engine was about $30,000 each. If you take the Consumer Price Index and the ratio of that – just taking cost of living adjustments – it would be up to about a factor of six. I'm absolutely sure they would have liked to produce 2,000 more. I'm pretty well convinced they would have been just as happy as clams to make another couple thousand of these things. We made about 650 engines. That was 650 production units. We actually flew 469 of them during the Apollo Program, which is an astounding number of little rocket engines that actually fly in space all on one program. In all that time, and the millions of cycles that were put on during that whole program, there was not one R-4D valve or engine failure. We were really kind of tickled that we may have done something good. (See Slides 14 and 15, Appendix E)

What happened to the other 200 engines? NASA had this requirement that you had to keep both the hardware and the documentation around for at least ten years after the conclusion of the program. We had a storage spot. We tested the original lunar module qualification unit with the four engines, which on both the service module and the lunar module were stored structures. That unit sat in the back of a storage room for about forty years before the company downsized; they were throwing stuff away, and Carl Stechman managed to recover one. I know where one other one is for sure – in my office. A bunch of them were sold to Rocketdyne for some classified program. They sold eleven of them for another program. NASA's Jet Propulsion Laboratory got a bunch of them, primarily to be taken apart and used for valve testing. There are some in the Smithsonian, a few in the space museum down in Alamogordo, New Mexico. There are some in NASA's Kennedy Space Center museum. There are some in other museums somewhere in the South. There is some facility that has quite a number still sitting in the original containers. Last I heard, there were still a number of them in storage down at Johnson Space Center, waiting to find a home. (See Slide 16, Appendix E)

> In all that time, and the millions of cycles that were put on during that whole program, there was not one R-4D valve or engine failure.

I was running a 200-pound thrust development program at Marquardt in about 1992. I actually took one of those engines off a particular module because we were having a data problem in our test stand. We couldn't seem to get data to repeat, or we couldn't get anyone to believe that we could repeat data anyway. We actually pulled a logbook for one of those engines out of its original dead storage, took one of those engines off of that module, put it into the test stand, and re-fired the thing. It repeated its original acceptance test within two seconds of specific impulse and about one-half pound of thrust. They seemed to have stood the test of time fairly well. There was some documentation still around up until a couple of years ago; I had a copy of the original PFRT report and the qualification report I wrote. Carl Stechman is a good source because he kept a lot of the original stuff from a historical standpoint. The thing has really gone a long way from where it started. It started as an injector. The basic injector was the R-4D that is still sold today. It was the same one that we developed just before the pre-igniter was incorporated because we got some thermal control. We got a little smarter. We turned that one with a large area nozzle made out of niobium into one with a 311-second engine. I left Marquardt at about that time. Then, they decided, "Gee, if a little is good, more must be better." It had more nozzles and some reductions in film cooling. (See Slide 17, Appendix E)

Today, the engine marketed by Aerojet is still that basic eight-on-eight configuration with the changes that they have made into it. It has a current specific impulse of about 323 seconds. From where it started out in the good old days at 292 seconds to where it is today, it's come a long way. (See Slide 18, Appendix E)

Gerald R. Pfeifer, Aerojet - Attitude Control Engines

Editor's Note: *The following information reflects a question-and-answer session held after Pfeifer's presentation.*

QUESTION: Give us a sense, if you would, of where all these different articles were tested – the ones that were to fly as well as the ones that were part of the development activity. Were there multiple facilities up and down the West Coast or were they all contractor? Were they altitude as well as ambient?

PFEIFER: All the engines were both qualification and acceptance tested at Marquardt's facilities. After we won the Apollo Program contract, we went off and built two vacuum test facilities, which simulated altitude continuous firing for as long as we wanted to run an engine. They would run days and days with the same capability we had on steam ejection. We did all of the testing in both for the qualification and the acceptance test. One of them was a large ball, which was an eighteen-foot diameter sphere, evacuated again with a big steam ejector system that could be used for system testing; that's where we did the Lunar Excursion Module testing. We put the whole cluster in there and tested the entire cluster at the simulated altitude conditions. The lowest altitude we tested at – typically an acceptance test – was 105,000 feet simulated altitude. The big ball – because people were interested in what they called goop formation, which is an unburned hydrazine product migrating to cold surfaces on different parts of spacecraft – was built to address those kinds of issues. We ran long-life tests in a simulated space environment with the entire inside of the test cell around the test article, liquid nitrogen cooled, so it could act as getter for any of the exhaust products. That particular facility could pull down to about 350,000 feet (atmosphere) equivalent altitude, which was pushing pretty close to the thermodynamic triple point of the MMH. It was a good test facility. Those facilities are no longer there. When the guys at Marquardt sold the company to what eventually became part of Aerojet, all those test facilities were cut off at the roots. I think they have a movie studio there at this point. That part of it is truly not recoverable, but it did some excellent high-altitude, space-equivalent testing at the time.

STEVE FISHER: Regarding your facilities, and probably more so towards the end of Marquardt, how did you guys test in the San Fernando Valley with all the leaks and the stuff you had? How did you guys manage the Environmental Protection Agency (EPA)?

PFEIFER: Surprisingly, we had very few problems while testing in the San Fernando Valley. In the early 1960s, nobody had ever seen dinitrogen tetroxide (N_2O_4), so that wasn't too big a deal. We really did only make small, red clouds. In all the hundreds of thousands of tests and probably well over one million firings that I was around that place for, in all that thirty-something years, we had a total of one serious injury associated with rocket engine testing and propellants. Because we were trying to figure out what propellants would really be good, we tried all of the fun stuff like the carbon tetrafluoride, chlorine pentafluoride, and pure fluorine. The materials knowledge wasn't all that great at the time. On one test, the fluorine we had didn't react well with the copper they were using for tubing, and it managed to cause another *unscheduled disassembly* of the facility. It was very serious. It's like one of those Korean War stories. The technician happened to be walking past the test facility when it decided to blow itself up. A piece of copper tubing pierced one cheek and came out the other. That was the only serious accident in all of the engines handled in all those years.

Now, we did have a problem with the EPA later because they figured out what the brown clouds were about. We built a whole bunch of exhaust mitigation scrubbers to take care of engine testing in the daytime. In general, we operated the big shuttle (RCS) engine, the 870-pounder at nominal conditions; they scrubbed the effluents pretty well. If you operated that same 870-pound force engine at a level where you get a lot of excess oxidizer, yeah, there's a brown cloud. But, you know, it doesn't show up well in the dark. They did do some of that. But, that's gone; it was addressed one way or another.

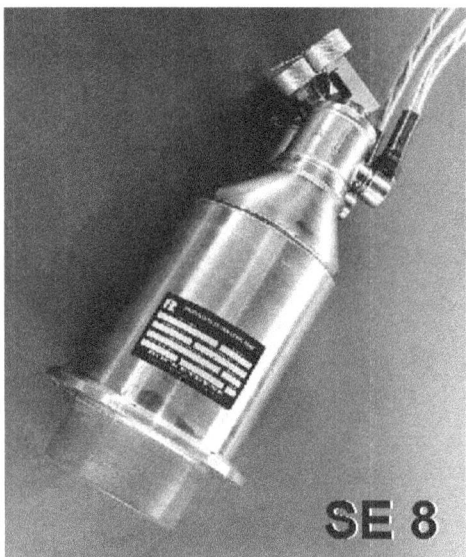

Saturn S IV B Stage Ullage & Apollo Command Module Reaction Control Engines (Source: Tim Harmon's presentation charts, Appendix F)

Chapter Four

Tim Harmon
Rocketdyne - SE-7 & SE-8 Engines

Tim Harmon has more than forty-one years of experience with The Boeing Company and its Rocketdyne Division. He retired as the chief systems engineer on the MB-XX Program, a cryogenic upper stage propulsion system. Harmon also was involved in ten Rocketdyne engine development campaigns, ranging from small attitude control engines to a large Aerospike engine. In the Apollo Program, he worked on development of the lunar module ascent and command module attitude control engines. Harmon earned a bachelor of science degree in aeronautical engineering from Purdue University, a master of science in mechanical engineering from the University of Southern California, and a master of business administration from Pepperdine University.

I hired in at The Boeing Company's Rocketdyne Division in 1963. There are two themes I would like to emphasize about my experience there: the environment at Rocketdyne, and using lessons learned.

It started out in 1957 with the first Sputnik launched into space. The United States' response was a vehicle that went all of four feet off the ground before exploding. Then, in 1961, the Soviets put an astronaut in orbit. What was our response? We got Alan Shepard up in space for fifteen minutes. When I started work at Rocketdyne, there was a space race, and we were behind. We didn't look that good. That kind of environment was intense. Rocketdyne's response was to hire more staff. We had employee serial numbers in those days. I hired on in April, and my serial number was 318083. By the end of the summer, there were many new hires. The staff I talked to that hired in June, July, and August had serial numbers that were a thousand points higher than mine. It was a good environment from the standpoint that it was growing.

When something is growing, there is opportunity, and it's a fun environment. On the other hand, in my first six months at Rocketdyne, I had three managers because they were getting promoted, moved, or changed. That was a little disconcerting. You have pluses and minuses with a growing environment.

Now, let's talk about off-the-shelf technology and lessons learned with the Apollo Program. There were a dozen SE-8 engines in the Apollo capsule; in the S-IV-B Stage, there were two SE-7 engines. These engines were similar in design and function. They were based on the Gemini Program vehicle attitude control engines. The Gemini Program used small engines that Rocketdyne built. These were buried engines. One of the technical issues with buried engines is keeping engine heat from soaking into the Gemini capsule. The solution was ablative-cooled engines that insulate the vehicle from engine-firing heat. The Gemini Program success was the basis that Rocketdyne used to win the contract to develop the engines for the Apollo Program's Saturn vehicle. (See Slide 2, Appendix F)

On the Gemini vehicle, we had orbital maneuvering engines (OMEs), and we had reaction control system (RCS) engines. The OMEs were 100-pounders (generating 100 pounds-force of thrust). The RCS engines were twenty-five-pound-force (lbf), buried systems, and depending on the angle, the nozzles were scarfed (truncated at an angle), so they wouldn't impinge on the vehicle. Rocketdyne said, "Well, let's go with an off-the-shelf design where we and you (NASA) can save money." It was a good idea, and I think it proved itself. (See Slide 3, Appendix F)

Gemini's 100-pound force Orbital Attitude Maneuvering System (OAMS) illustrates some of the problems we had. We had two propellant valves. We had an ablative thrust chamber with different wraps of ablative material. There were a silicon carbide liner and a silicon carbide throat. One of the problems with silicon carbide material, while very sturdy, was that it would crack due to thermal shock. The thermal shock issue was never a problem on the Gemini vehicles. The throats would crack and the sleeves would crack, but they would all stay in place. That was the design, and NASA went with it. It was man-rated (deemed safe to carry humans) and successfully flew. The Gemini thruster performance was adequate. As a pressure-fed system, it was fairly simple. The S-IV-B engine has the same look as the Gemini engine. They did de-rate the thrust level. NASA didn't feel they needed the thrust. It also was pressure fed (no propellant pumps). It had a 274-second specific impulse using nitrogen tetroxide and hydrazine propellants. The design year was 1965. This engine application, for the S-IV-B Stage, was to settle propellants into the bottom of the stage, so they could start and restart the J-2 engine. It was a continuous burn, with no engine pulsing. This fairly simple engine was a "lessons learned" direct application, and there were no development problems. (See Slides 4 and 5, Appendix F)

The 70-pound SE-7 engine is very similar with its two valves, ablative material, a silicon carbide liner, a silicon carbide throat, and overall configuration. There were different wraps. One had a ninety-degree ablative material orientation. That is important because it caused problems with the SE-8, but not for this application. It was not overly stressed. It was a validation of the off-the-shelf application approach. There were two SE-7 engines located on the stage near the bottom. They had their own propellant tanks. That was the application. All it did was give a little bit of gravity by firing to push the propellants to the bottom of the tanks for start or restart. It was not a particularly complicated setup. (See Slides 6 and 7, Appendix F)

What had we learned? This was a proven engine in a space environment. There weren't any development issues. Off-the-shelf seemed to work. There were no operational issues, which made the SE-7 very cost-effective. Besides NASA, the customer for this application was the Douglas Aircraft Company. Douglas decided the off-the-shelf idea was cost-effective. With the Gemini Program, the company was McDonnell Aircraft Corporation, which was part of the reason the off-the-shelf idea was applied to the Apollo. (See Slide 8, Appendix F)

However, here are some differences between Apollo and Gemini vehicles. For one thing, the Apollo vehicle was really moving at high speed when it re-entered the atmosphere. Instead of a mere 17,000 miles per hour, it was going 24,000 miles per hour. That meant the heat load was four times as high on the Apollo vehicle as on the Gemini craft. Things were vibrating a little more. We had two redundant systems. Apollo was redundant where it could be as much as possible. That was really a keystone or maybe an anchor point for Apollo. We decided to pursue the off-the-shelf approach. However, the prime contractor was a different entity – the North American Space Division. They thought they ought to tune up this off-the-shelf setup. It was a similar off-the-shelf application, but at a higher speed. They wanted to improve it. What they wanted to improve was the material performance of silicon carbide. They were uncomfortable with the cracks they were seeing. They were uncomfortable with the cracks in the throat, and feeling that the environment was a little tougher, that maybe it was going to rattle, perhaps something would fall out, and they would have a problem. They wanted to eliminate the ceramic liner, and they wanted a different throat material. (See Slides 9 and 10, Appendix F)

The Rocketdyne solutions were to replace silicon carbide material with a more forgiving ceramic material. Also, due to the multiple locations within the vehicle, the shape of the nozzles varied. Some nozzles were long, and some nozzles were short. We came up with a single engine design with variable nozzle extensions and configurations to fit particular vehicle locations. (See Slides 10 and 11, Appendix F)

> Instead of a mere 17,000 miles per hour, it was going 24,000 miles per hour. That meant the heat load was four times as high on the Apollo vehicle as on the Gemini craft.

Chapter Four

> We came up with a single engine design with variable nozzle extensions and configurations to fit particular vehicle locations.

The engines cost about $35,000 in 1963, which was how much you paid for a Rolls-Royce automobile. So, there were twelve Rolls-Royces on every Apollo vehicle. That's important because it meant we really had a fair amount of investment by the time we got to the end product, and we didn't want some problem to show up that would cause us to discard the engine.

Another kind of issue was the work environment. There was a lot of competition at Rocketdyne for test site slots. We had a lot of simultaneous programs in the works of large engines and small engines. Test slots were an issue or a competition between various groups. There was another competition going on as well. It was organizational, between the quality and manufacturing departments. The environment was such that quality wanted to catch engineering oversights; or manufacturing wanted to catch engineering oversights. It was friendly, because we were all trying to win the race to the moon, but it was still a different environment.

I was the test engineer during engine development testing, and had to cope with the space race and its schedule pressure. I was assigned to assemble an SE-8 engine with a nozzle to get it ready to test. We accidentally installed an O-ring made of the wrong material. I tested the engine, and the seal leaked post-test. My name was Mudd. My boss, Cliff Hauenstein, who was the materials engineer, had to explain this to the vice president at Rocketdyne. This Rocketdyne vice president was a tough, old-school manager. He looked like Yul Brynner[1], and his management style was like Attila the Hun. It was not going to be a comfortable meeting. The vice president had jelly beans on his desk. He always ate jelly beans. He was a nervous guy. We walked into the office, and a jelly bean caught in his throat. He ran out, turning purple before the meeting started. Finally, he came back in, having coughed and dumped the jelly bean. But, we felt he must have seen God because he was extremely mellow. Everything was fine. "Don't let it happen again," he said. After that, my manager always had jelly beans in his office. It's a true story, but the lesson learned was, "You know, these things are expensive. You ought to have a little better control and awareness, so you don't make a mistake like that."

We also had an uneven throat erosion issue caused by uneven injector performance. When we went away from the silicon carbide throat material, the substitute material we chose was not as sturdy. The SE-8 program was still schedule driven. We could identify injectors that caused high erosion problems by hot-fire testing each injector in a specially built thrust chamber that used steel throat inserts. A poorly performing injector would erode the steel throats quickly. The plan was to devise some sort of tool for inspecting throat erosion post-test to identify samples with good erosion characteristics and samples with bad erosion characteristics.

---

[1] Accomplished American actor who played several roles requiring a serious demeanor.

By this time, we probably had a database of fifty or so injector tests on steel throats of various discolorations or erosion patterns. I made up a board that basically consisted of samples that were good and samples that were bad. We put them on a metal strip. To do that, and do it in a hurry because we were in production, I went into the shop, found a welder, and said, "This is a 'G job.'" In those days, a G job was special. It took no paperwork, and you could get anything done you wanted. So, we would weld a throat on a strip of metal. Now, it was hot. You had to wait for the first weld to cool down before you could weld the next one. Again, we had injectors at the Rocketdyne test site we were trying to inspect and show were acceptable. This tool was needed in a hurry. We would weld one on, then wait for the thing to cool. It was time-consuming, and I needed to solve the problem. There was a men's bathroom right next to welding station. The welder would weld one strip, and I would run and flush it in the urinal and cool the metal weld immediately. We would then complete the next weld, flush it in the urinal, cool it right down, and repeat the process. (See Slides 12 and 13, Appendix F)

I had this nice board, all these welded pieces, delivered to the test site. I brought the tool out to inspection, and they were really happy with it. We fired the next injector. An inspector looked at it and said, "It feels all right to me. It looks just like this one on the inspection board." However, the quality inspectors were unaware that the samples had been in the bathroom being flushed. I thought that was kind of payback for all the grief I got from the inspectors.

Another issue we had was the appearance of the thrust chamber post-test. This ablative material looked like a piece of polished wood before hot-fire testing. After we hot-fired it, it had ablated, so it looked a little rough, with something like the appearance of burned wood or charcoal. After one hot-fire test, the Air Force inspector said, "Well, it doesn't look like this nice piece of wood, and we don't know whether to accept it or not because it looks worn." The solution to that was elegant, but silly. We changed the chamber inner diameter appearance to one that looked like it had been fired. We took this piece of the inner sleeve, and we charred it in an oven, so that it looked already used. Then, when it was hot-fired, it didn't change in appearance and the inspectors thought that was neat. Now, that kind of fix for an issue was a waste of money. Everybody was happy, but it was an added step. Eventually, we convinced them that it was a stupid idea, and this fix was eliminated. That was sort of humorous, but it also was a sign that people were not working together or were not understanding the processes. Who was at fault? Nearly everyone – engineering, inspection, NASA, and the customer.

Those were two big issues and their solutions.

Our test facility, Santa Susana (California), had an interesting history. A lot of B Western movies were made there before it became a test facility. If you see a 1920s or 1930s Western, there was a good chance it had been filmed at Santa Susana. Rocketdyne purchased the area for testing rocket engines and built many facilities. This was the facility that tested most small engines. It had four or five small engine test stands. It was a very active facility. Engineers were pushing for their test slots and trying to get things done. We had to test our engine at altitude conditions. We had a very limited capability in terms of altitude simulation testing. It could run approximately fifteen minutes before needing to be reloaded with propellants. To hot-fire calibrate an engine or run development tests, we conducted a test, analyzed our data, and determined our next test condition. If we wanted to change the mixture ratio and the various tank pressures to get another test, we tried to do that as fast as we could with our handy, dandy slide rule calculators. Times were different for testing in the 1960s than they are now. It was fairly frantic because we tried to get as many test conditions as possible in the shortest time span. What we did have were Datelink Independent Gateway Retrofit (DIGR) data recorders - we called them diggers - for instrumentation, which was interesting. The DIGR was a circular plot with an ink pen that showed the test data. A development engineer would run a test, go over and look at the DIGR, and it would tell the value of pressure or thrust or whatever. But, it wouldn't tell right away because there was a hysteresis (or lag time) in the pen. We would have to go over and tap the DIGR to get our values. As soon as we ran a test – we had probably ten DIGRs we had to pay attention to – we were going along tapping every DIGR to get our value. It looked like we were doing some sort of comedy because we would tap all the DIGR, use our slide rules, do all the calculations, and then yell new test values to the test conductor. We would yell, "Change the tank pressures to this." Then, we would do the next test, and start all over again. It was very frantic and very challenging. (See Slides 14 and 15, Appendix F)

When I was training to be a development engineer, I was mentored by a very seasoned engineer who had hired in six months earlier than I had been. We worked together for about two weeks, then I was on my own for running tests. The very first test we ran, the engine split in half. I could not believe it. I had a guilt trip that I had just ruined the space program. I got over it, but it was still traumatic. However, in those days, failures were somewhat acceptable because we were trying to develop engines, and it was exciting to be part of the space race.

Development programs generated a huge amount of data. We had a lot of people doing tests. If one person wrote the test report on an engine, that was fine. That was not a problem. However, when you had twenty engineers conducting tests and writing reports, there was no consistency. What one engineer thought of the engine, another engineer would have a different slant. That turned out to be a major issue. *The solution was to develop a simplified reporting process, a template system for reporting processes.* (See Slide 16, Appendix F)

Engine development issues were generally resolved through an extensive test program. That's a lesson learned. For the next Apollo-type program, I think you have to be able to anticipate issues. Every application is slightly different, and the customer is slightly different. The customer wants to improve the engine. Many of those improvements cost development activities that are not necessary. It's hard to say, but we certainly had an engine that was developed and worked. However, you have to anticipate there will be problems. I think that is the same situation you will find with the J-2X upper stage engine development program. That application is slightly different. Nothing goes quite off-the-shelf, but it's better than starting from scratch.

In summary, Apollo in the early days was certainly a challenge. The challenge was the fact that we were behind. As a country, we didn't look nearly as smart technically as the Soviets. That was a challenge, and that was a driver. Using the off-the-shelf approach saved time. It leveraged previous technology. However, we also needed to work together. The goal was the same for engineering, for quality, for inspection, for the customer, and for NASA – work together.

Editor's Note: *There were no questions taken at this time. The question-and-answer session was held after Mr. Harmon's second presentation, the transcript of which appears in Chapter 7.*

AJ10 Service Propulsion Engine (Source: Clay Boyce's Presentation Viewgraphs, Appendix G)

Chapter Five

Clay Boyce
Aerojet - AJ10-137 Apollo Service Module Engine

Clay Boyce left the U.S. Air Force in 1955, and joined Aerojet at its Azusa, California, facility. After the launch of Russia's Sputnik Program, he joined the Thor/Able Second Stage Program, an adaptation of the Vanguard second stage. Subsequently, he participated in the conceptual design and development of the AbleStar upper stage. Variations of this stage are still flying. In 1960, he was assigned to the Win Apollo team and, then, became engineering manager of the Apollo Service Propulsion System (SPS) engine development. In 1969, he was assigned to NASA's Johnson Space Center in Texas for technical support of the SPS engine. Later assignments included the Space Shuttle Orbital Maneuvering System engine, the Japanese N-2 Upper Stage, and the National Aerospace Plane. Boyce retired from Aerojet in 1991, and has provided consulting services to the company since that time.

Some of the most tense times I ever had were on the Apollo mission. The question I always got asked the most when somebody talked about it was, "What if the service module engine doesn't start, especially when the astronauts are ready to come home?" I will tell you why it was tense. I knew it would start, but until that first mid-course firing on the way to the moon, I didn't know if they had connected it right. Once I got a firing signal, I knew they had mated the connectors properly in place.

Our Apollo proposal effort started for me in 1960. The Thor AbleStar vehicle was being developed then. (See Slide 3, Appendix G) We used technology like the titanium nozzle for the

Chapter Five

SPS engine. I was partying in a swimming pool down in Florida just after the second Able-Star launch when they came and got me about 3 a.m. and said, "We booked you a flight out of Orlando, 7 a.m. Be in Philadelphia, go to General Electric, find the Apollo group, and tell them our engine is the best." That was my first introduction to the word "Apollo" in the space program.

We had written twenty-nine proposals to twelve different primes (contractors competing for Apollo), and all but one of them selected our engine. We had a pretty good chance of winning the job. The engine was the same height as a space shuttle main engine, and the exit diameter was slightly larger than the space shuttle main engine. (See Slide 5, Appendix G) The size was set before Apollo adopted the Lunar Excursion Module (LEM) mission approach. Prior to the LEM, the whole Command Module was to be landed on the moon, and this engine would have been the engine to lift it off the moon and return it to Earth. I would have had to wait a long time for that first firing, if they had followed the original concept. Aerojet was awarded the Apollo SPS engine development contract in April 1962, and the final decision on LEM was made about November or December of that year. We had proceeded far enough that even though they went to the LEM concept, they decided it wasn't worth starting over with a smaller, lower-thrust engine, so they kept the large size. The large size resulted because it was specified to be a pressure-fed engine. They didn't want to have to worry about the reliability of pumps and everything on the moon at that time. It was a 20,000-pound-force (lbf) thrust engine, but it was as big as a space shuttle main engine (SSME). Chamber pressure was only 100 pounds per square inch (psi). The major change that it made from the mockup was at the head-end, gimbaled configuration was changed to a throat-gimbaled design, which reduced the overall height of the Saturn vehicle when they put the LEM in behind the service module. The throat gimbaling of the engine saved about three and one-half to four feet of overall vehicle length and weight.

The technology we had at that time was the Saint/Apollo subscale engine. (See Slide 4, Appendix G) It was for a satellite interceptor system for the Air Force. It used what we called earth-storable propellants, N_2O_4 (dinitrogen tetroxide, a powerful oxidizer) and some of the hydrazine families, because the Saint Program actually was an in-orbit satellite that was supposed to be able to detect an orbiting enemy satellite and go after it. That program was started in 1957-1958. It never flew because the computer capabilities and the control system it required just weren't mature enough to do it. But, we did build the ablative engine, the biggest at that time with 2,000 pounds of thrust. It had a nice titanium nozzle, an aluminum injector, and an ablative thrust chamber–all the things we needed for the SPS because we were looking for the simplest, lightest-weight engine we could get. We later used that engine for subscale testing of ablative materials and nozzle extensions for the full-scale engine.

The general configuration of the SPS engine was 20,000 pounds of thrust, with a chamber pressure of 100 psi and specific impulse (Isp) of 314.5. The very large nozzle had an area ratio of 62.5:1 (exit area to throat area). The propellants were nitrogen tetroxide (also known as N_2O_4 and nitrous oxide) and A-50. A-50 was a hydrazine family fuel. Aerojet developed it for the Titan Missile Program when they went with Titan II, to store it in the launch silos. They wanted the highest performance they could get. N_2H_4 was just pure hydrazine, which doesn't take low temperature very well. In fact, it freezes about like water. We started adding unsymmetricaldimethylhydrazine (UDMH) to the hydrazine until such time as it would meet the environmental specifications the Air Force needed for Titan II. It turned out it's roughly a fifty-fifty mix. We still had to be careful with that fuel because the two fluids didn't mix very well chemically. We had to spray the two fluids through some special nozzles to get them to emulsify with each other into a single fluid. If we ever got it too cold or froze it, the hydrazine separated back out. Then, if we tried to run the engine, things could go boom in the night.

The inlet pressure was only 165 pounds per square inch absolute (psia), but we needed at least forty psi pressure drop across the injector just to get some kind of stable flow. It was a whole new game for some of us. We didn't have much supply pressure to work with. It had the aluminum injector to keep the weight down. That was a couple feet in diameter, and we didn't have a lot of propellant to cool it. In fact, we had to use both propellants to keep the injector cool. There were twenty-two ring channels in the injector. Specification required 750 seconds duration, or fifty engine restarts during a flight.

There were several first flight things we accomplished with the engine. It was the first ablative thrust chamber of any size to fly. (See Slide 6, Appendix G) There were no liners in it. It was just straight ablative material. It took us a while to figure that out. It was a throat-gimbaled engine, and it was the first engine to fly with columbium (also known as niobium, used as an alloying element in steels and superalloys) in the nozzle.

The first time we fired that assembly, we got a couple of surprises. (See Slide 7, Appendix G) The chamber pressure was supposed to be 100 psi. We fired it up, and it ran just fine for about a second, then the chamber pressure (Pc) dipped down to forty psi and gradually came back up to 100. The engine went on running, but we were holding our breath a little. One of our requirements was to use the fuel to actuate the propellant valve. With those low pressures, it pulled so much fuel from the inlet of the engine to actuate the valve that there wasn't enough to run the engine for a second or two. The good news was it stayed stable through all of that. During these very first firings, we hadn't received the requirement for dynamic stability, so the engine contained an unbaffled injector, and it recovered. It never really ever went unstable. It just turns out it was fairly throttleable.

> It was a throat-gimbaled engine, and it was the first engine to fly with columbium in the nozzle.

There was another unexpected surprise that involved the interface with the actual Service Module. The decision for our approach evolved during proposal time. Half of the contractors looked at a single engine with redundant controls; others had groups of four and five engines for reliability. (See Slide 8, Appendix G) The final contractual decision, from a weight and an efficiency standpoint, was to make a single thrust chamber with dual redundant valving to assure that we could always get a start, and always get a shutdown. The gimbal actuators, which were the other moving parts on the engine, had semi-redundancy. The propellant valve was about two feet square and weighed 100 pounds, the biggest valve on which I ever worked. It was kind of a mirror image. One side of it was the oxidizer; the other side was the fuel. There were four ball valves in each propellant circuit in this arrangement: a fuel ball, a common shaft, and an oxidizer ball. The actuator to rotate the shaft was located between the balls. There were four of these individual assemblies utilized into a series-parallel configuration. On an engine start, any one of those actuators could fail, and the engine would still start. If, on shutdown, one of the actuators failed, it would still shutdown. We could run the engine, which we normally did with all four sets of valves open, but it would run equally well on one pair in each circuit, which we did a lot of in qualification testing.

One of the development challenges we faced didn't have to do so much with the valve operation as it did with manufacturing. It was a complex casting with lots of machining. We were having them machined outside our plant by a subcontractor, and it was taking him about a week to make one casting. Somewhere in the middle of trying to get more castings, Aerojet purchased what, at that time, was one of the newest numerically controlled machines with multi-axis operations. Management decided they needed something to do with it, so I got directed to bring the machining of the valve in-house. We did that, and it took a couple of months to get the things all set up. We started machining valves. On the fourth day of the six-day cycle, two shifts a day, the machine tool would jam a tool through a side of the casting and ruin it. At that time, the machine control was not digital. There was a punched paper tape, about one inch wide and over 100 feet long. Every time they would ruin one of those castings, they would have to go through that whole tape, make a few changes, and punch out an entire new tape. It was three and one-half months before we started getting good castings. After that, it worked fine, but it did set us back a little on our schedule.

The other major valve problem was the actuation. We were required to use engine fuel as the hydraulic actuator fluid. They wanted to minimize the number of connections that crossed the interface between the Service Module and the engine. We were only allowed the main propellant lines and one redundant, electrical cable. That was it. We switched to a pneumatic valve actuation because we had to stay in the same envelope, and we couldn't cross the interface. We removed the fuel hydraulic system from the cylinders and put a big spring in there that was used to close the valve and an engine-located, high-pressure nitrogen subsystem to pneumatically open the valve. We were required fifty starts of the engine. The amount of nitrogen we had would restart the engine about eighty times. (See Slide 8, Appendix G)

Those were the two major events during valve development. Once we got them going, we had the usual problems of seal material in the valve. We started out with Teflon, and it would cold-flow too much. We finally used some fiberglass-impregnated Teflon materials that would work. We had a three-micron finish on those valve balls, and they were each about two and one-half inches in diameter. It was a job manufacturing them, but not intolerable. The balls were made out of stainless steel. During one of our weight reduction exercises, we got the bright idea to make the balls from some lighter-weight material. We tried beryllium, which was about the only candidate that was significantly lighter. They worked, except that beryllium is pretty soft, and it didn't have wear life that would meet the mission requirement.

Our injector assembly was about two feet in diameter. It was all aluminum. (See Slide 9, Appendix G) It had five baffles for the dynamic stability, and I think we did 240 or 250 bomb[1] tests in developing this baffle configuration. The original injector had twenty-two rings. There was one main rule in the fabrication of the injector: we had to keep parent material between the propellants. There couldn't be any weld that both propellants could touch. With the low pressures and the big surface area to cool, we didn't have much room left for lands between the rings. That was a problem. When we went to the baffles, we had to reduce the number of channels down to fifteen. When we started injector development, because of the narrow land area, we thought we would have a welding problem, so we decided we would braze the injector. Well, it turns out there were about 200 linear inches of area that had to be brazed. If we didn't have it perfectly cleaned and perfectly aligned, we didn't get a perfect braze. We never got a brazed injector that was 100 percent leak-free.

About that time, NASA was still looking at the future Nova launch vehicle, and Aerojet had the contract for the M-1 engine for the Nova's second stage. It was 1 million-plus-one-pound thrust hydrogen/oxygen engine. Aerojet had bought an electron beam welder for use on the program. When the M-1 got canceled, it sat in a corner, doing nothing. We decided we would give it a shot at welding the SPS injector, and it worked quite well. After we started firing the injectors for long periods of time, we encountered a problem that became a phenomenon in very low-pressure, large engines with the particular propellant combination we had. We named it "popping." We would start the engine up and we never knew when, but if we were watching the chamber pressure (Pc) trace on the oscillograph, all of a sudden, the trace would expand

---

[1] Here the 'bomb' refers to an impulse charge to induce instability in the combustion chamber for assessing the engine's susceptibility to combustion instability.

Chapter Five

> We had to worry more about damage to the injector from bomb fragments than from anything else.

into a little football-like bubble. Then, it would go away. It would just last a few milliseconds. But, every time it would make a bubble, to us in the test bay, it sounded like a "pop;" hence, the name "popping." Eventually, it became a large-enough concern that NASA convened a meeting at North American Aviation, Inc., and brought all the stability experts from the universities, and we spent two days talking about that. I thought, "Boy, I'm finally going to get some help." The next day, I got the results. They told me they didn't like the name "popping." We eventually solved it. At the low pressures and the low pressure drop across the injector, the length of the drilled orifice became critical, and the propellant stream sometimes would come out clean, and sometimes would attach to the edge of the orifice, causing some propellant to flow onto the injector face. When the propellant stream ran along the injector face, it changed the combustion characteristics of the element, and it would cause the pop. It never hurt the engine, but the first few times really worried us. We counter-drilled those and got a shorter, fixed-length orifice that solved the problem.

We had to do the bomb tests to check the dynamic stability. That turned out to be relatively easy to do. We had to worry more about damage to the injector from bomb fragments than from anything else. The ablative chamber had no kind of liners or anything inside; it consisted only of laid-up ablative material with some fluted aluminum flanges at each end. (See Slide 10, Appendix G) At the time we started the program, and even back in the old Saint Program, we didn't have nice impregnated tape like they do today. The coated glass fiber string material was made in whatever length rolls you wanted. The continuous string was a little bit bigger around than a piece of heavy kite string. Whatever thickness chamber we wanted, we'd cut that length of strings and lay them on a piece of sticky tape. All of them were hand laid, side-by-side-by-side. The tape with the strings was then wound around the chamber mandrel (a spindle to support the piece during machining) and impregnated. We did that until the ablative material industry progressed and developed the flat tape. The flat tape made the chamber lighter because the ratio of the glass to resin was a lot greater, and provided more heat-dissipating capacity. This allowed the chamber wall to be thinner.

The next major component on the engine was the gimbal actuator. This particular actuator is a semi-redundant device. I say "semi" because the main actuator movement device was a single ball screw, much like that used in present-day car steering mechanisms. The ball screw was operated by two electric motors transmitting power through magnetic particle clutches. One motor/clutch assembly extended the actuator, and one retracted the actuator. There was a completely redundant set of motors and clutches in each actuator for mission reliability. During engine operation, all four motor/clutch assemblies were running, and control signals from the spacecraft activated a specific magnetic clutch for the desired engine position. A couple of our problems with the actuator didn't have anything to do with technical accomplishment. At the time we got the contract for the SPS engine, we planned to subcontract the actuator to a company called Lear Inc. in Grand Rapids, Michigan. Lear was the biggest actuator company in the U.S. at that time; all Boeing and McDonnell Douglas Aircraft Company airplanes used Lear actuators for moving controls surfaces and flaps. They had the best reputation for that. We gave them a contract, and they went to work.

The actuator assembly had to be enclosed in a pressurized, hermetically sealed can. (See Slide 11, Appendix G) The electric motors and the magnetic clutches wouldn't work in the vacuum of space. After a year or so, we received a pair of actuators to start testing. Both units were overweight and larger than the required envelope. Right about that time, another company by the name of Siegler bought Lear. The Lear group kept working the size and weight condition, along with a stiffness problem. Then, Siegler bought another company called Jack and Heintz Motor in Cleveland. Well, the Lear plant was overloaded with work; they didn't want to build a new building, and the Jack and Heintz plant in Cleveland had lots of empty space. Lear/Siegler moved the actuator program to Cleveland. This was a problem because the actuator engineers wouldn't move to Cleveland. All of a sudden, we had this contract with about half the money spent, but we didn't have a part, and we didn't have an engineering staff anymore. About that time, North American Aviation and NASA were getting nervous enough that they decided we'd better find another actuator company. Meanwhile, a couple of people who had worked at Lear had gone to work for an actuator company down in Costa Mesa, California, called Cadillac Gage. They were building actuators for the nozzles on the Polaris missile under contract from Aerojet, and had produced good hardware. The ex-Lear people talked to the Cadillac Gage management, and they gave us a bid to build the actuator. At that point, NASA and North American interceded and said, "We have already spent half the money. We can't afford that again. You are directed to give a fixed-price contract." We gave them a fixed-price contract, and they started to work.

Then, the second problem we found on the Lear actuators resurfaced. The stiffness requirement for the actuator was to be a spring rate of 300,000 inch-pounds to meet the vehicle control system loop frequency. The stiffness on the two actuators we'd received from the old Lear Company and the first two from Cadillac Gage were well below requirements. One of them was about 220,000 inch-pounds; the other one was 160,000. By then, the second company had used up all their fixed-price money and a little bit more. They said, "Sorry, but we have to stop work." It became a contractual battle for a while. It eventually was resolved. About that time, back at NASA Headquarters in Washington, D.C., Joe Shea, then the chief engineer of the Apollo Program, was reviewing his overall Program Evaluation and Review Technique (PERT) chart, and the actuator popped up as the umbrella for the whole Apollo Program, because if we didn't have that, the vehicles didn't go. Shea came to visit us one day. We went through everything with him. The North American guys were there and everything. After the review, they all went back to North American and said they would call us with their recommendations.

Chapter Five

> We were scratching our heads about what to do when one of those fortuitous events occurred.

The next afternoon I got a call that said, "Don't bother to come down. The problem is solved." I said, "What happened?" My 300,000-pound actuator was being tied to a 90,000 inch-pound spring rate bracket on the Service Module. It didn't make any difference as long as my spring rate was over 100. By then, we'd worked for almost two and one-half years trying to get that thing stiff, so we hadn't performed life testing on the motors and magnetic particle clutches. The clutches became the next challenge. There had never been a magnetic clutch that big. Existing technology were little things, less than three-quarter inches in diameter. This one was not quite three inches in diameter. As you might expect, we found we had overheating and other problems. We got it all done, but the problem illustrates the fact that sometimes you run into events that have nothing to do with your engineering capability. My final challenge out of the actuator development was the boss saying, "Okay, go get out of that fixed-price contract." If you've read government contracting regulations, you know there's no way you could do that. I had to figure out a whole different contracting scheme so we could get Cadillac Gage back to work. (See Slide 11, Appendix G)

The first nozzle for the Service Propulsion System engine was developed out of Saint and AbleStar technology. The nozzle was made out of titanium. We calculated that the area ratio at the head end was about 6:1 as the point where exit temperatures would be down to titanium capabilities for radiation cooling. The exit area ratio was 62:1. The first problem came with fabrication of the gores. It takes sixteen gores to make one nozzle. Our nozzle fabrication contractor's plan was to hot stretch-form the titanium, make a mold of the shape we needed, and use Calrod® units (elongated heating elements) to make it hot. That, in turn, would heat the titanium stretched over it. That didn't work too well. We were using titanium for the nozzle because it took lots of temperature. We had to get it very hot to stretch-form it. Titanium has a tremendous memory. It doesn't want to stay in a new place. It kept wanting to go back to its original flat shape when it cooled. We were in a quandary about that. We had only been able to make four gores that were even halfway close. We gave them to our subcontractor who was developing the technique for welding them all together. We were scratching our heads about what to do when one of those fortuitous events occurred. (See Slide 12, Appendix G)

About the same time we started scratching our heads, the then-president of our company was at a cocktail party in Washington, D.C., and he got to talking with the good lady senator from Maine. She said, "Dan, I gave you a bunch of money, supported you for that Apollo thing, but you're not putting any of that work in my state." I then got a memo that said, "Go see if they do anything in Maine that'll help your engine program along." I put all my problems in my briefcase and went and found a little guy in Maine who worried about the state's economic development. He listened to me for about half a day and said, "We've got nothing like that up here. But, two weeks ago, there was a guy in here wanting us to help him start a small fishing boat business. He wants to make them out of aluminum, and they're all nice and round, instead of square. He had a machine that makes stuff kind of round and long like you're describing." I got the boat maker's name and an address, which was down in Massachusetts.

When I got to the address, it was at a road crossing with a little building about the size of an old outhouse sitting there and that was it. Finally, about a mile away, I found an old cotton mill from back in the textile days, and here was this little, old Norwegian toolmaker, and he had built this big machine. He was making boat parts, specifically the double-curved part of the bow. Well, he hadn't been able to sell the boat business, so he had taken a couple of contracts to make gores for radar antennas.

Those gores, curved two ways, weren't too different from what was needed for the nozzle. I described what I needed and he said, "Yeah, I think I can do that. Send me a mold and send me some material." We did. Probably a month or six weeks later, he called and said, "Come on back. I'll show you what's going on." We went back to his shop in Massachusetts, and he had been working with his machine, and he had made a bunch of gores out of aluminum while he was working. So, I was going to be there for the first titanium pull. When we had tried the hot stretch-forming, we had to work with one big sheet of titanium. The boat maker's new process worked so efficiently, he could cut the sheet diagonally and we could get two gores out of every sheet. He laid the sheet out on his machine, brushed on a lubricant from a nearby pot sitting on a hot plate, and started the machine. It was a cold-working process, and the machine put that titanium through something like old washing machine washer rollers. He could vary the axis of the rollers and their distance apart. The machine put that titanium through a little S curve about two inches high; it took away all its old form memory and put in a new one.

The very first one that rolled out the end of that machine looked perfect to me, better than any I had seen. We laid it on the mold, and he got a little hammer and went all over it. "Well, it's not right. There's one little place that sounds a little different," he said. He went over and tweaked his machine, put the same part back in there, ran it through again and had a perfect part. I said, "How many of those could you make a day?" He replied, "I can probably make twenty of them in a day." I said, "How much?" He said, "Fifty bucks apiece, if you supply the metal." That was the last gore I got from him for three and one-half months because his business didn't exactly meet NASA 200-2 quality specifications. I had to create a Quality Department, a Purchasing Department, all of the "isms" and documentation to certify the parts were useable for nozzle fabrication. But, by then, that problem was solved, so I didn't worry about it anymore.

We finally got a nozzle together, and because of the size and the pressure, the only place that you could hot-fire test it was in the big Air Force altitude facility at Tullahoma, Tennessee. I talked recently to a gentleman who works at Arnold Engineering Development Center, and we decided the tests were conducted at the facility's J-3 cell, a very tall structure, that was used. I found out later when we got the test cell operating satisfactorily, that firing the SPS engine caused the test stand to flex and caused misalignment of the performance measuring instrumentation. The thrust and side forces measuring cells had to be biased for the thrust stand movement so we could get accurate test data. Around early 1964, we got the first altitude engine all together and set up at Tullahoma. For the first engine firing, I was sitting in the control room with one of the North American Aviation guys. We heard the big countdown going on, and they finally said, "Fire." We were looking at our oscilloscope and nothing seemed to be happening. We thought, "Man, that engine runs awfully smooth." Then, somebody said, "Okay, fire that engine," and they threw the SPS start switch. It turned out the whole countdown was for the rocket engine injector they were developing to pull the tunnel down to altitude. *That* was the big thing to them. We were going to fire the SPS engine for fifty seconds as a checkout. We noticed a little decay in thrust level after ten to fifteen seconds. But, then, it leveled right off and everything was fine, and we thought we had a successful run because everything looked very good.

As expected, the nozzle got very hot. Things finally cooled down, and we were able to get in the cell. It had gotten hot all right – too hot. The whole nozzle had moved forward about eighteen inches. About ten inches down, the nozzle had curled right back over itself. We had a nice, big S curve all around and a nozzle that was a foot and one-half shorter. That's when we figured out we needed columbium. The only reason we found columbium is when titanium wouldn't work, we went to the periodic table of the chemistry book and said, "Okay, what's the next lightest metal?" We determined it needed to handle at least 2,100 degrees Farenheit. We were running around 1,900 degrees at the nozzle attach point, and columbium appeared to be capable. Then, we tried to find some columbium. It turns out the nuclear power industry used columbium to make the rods in which they put the nuclear material for power plants. We went to the company that was making those rods.

The company was called Wah Chang, in Albany, Oregon, and was owned by a very nice Chinese gentleman. We wanted thin material. All the material he made was one-half inch or more plates. He thought he could make us some thinner stuff. We wanted it forty-thousandths (0.040) of an inch thick. He was able to produce the thinner sheet stock we required. The titanium nozzle had a machine flange on the front. We tried to follow the same design. We were going to make a flange out of columbium also, but nobody had ever made anything like that out of columbium. They had always used the flat roll stock. But we found a company that would try to make a casting. They had a big, spin-casting rig. That flange was around twenty inches in diameter. After making a casting, it turned out they never had machined columbium, and it is extremely tough to machine. They could get about two cuts around the flange with one, good silicon carbide tool, then it was shot. We later got rid of the flange and figured out an attachment where we could just roll a lip on the end of the nozzle sheet material and clamp it to the end of the ablative chamber. While they were busy making sheet metal stock for the full-scale nozzle assemblies, we got a small piece of columbium and just rolled it up to make a small regular conical nozzle. We put that on the Apollo subscale engine and took it down to Tullahoma to see how the columbium worked. We fired a partial simulated SPS duty cycle. That was 200 seconds firing and one half-hour coast, then re-fire. Everything went fine until the re-fire, when the thrust level was not where it should have been. We opened up the tunnel, and the nozzle was gone about two inches down from the flange. It was just all jagged.

We determined that the temperatures to which we raised the columbium caused it to absorb hydrogen from what little atmosphere there was in test cell. The hydrogen crystallized the columbium. The inside of the nozzle, next to the fire, was clean and nice–no penetration there. But, on the outside, you could see a crystalline surface. The engine re-starting shock had shattered the nozzle.

Operating the engine out in space would have been fine, no hydrogen. But, we had to develop and qualify the engine here on the ground. We had to find some way to protect the columbium so it could pass all the tests. We went through about twenty different coating materials and finally found one that North American Aircraft Division was using in some of the jet engines exit areas. We still had the same problem. We had to get the columbium up to 1,800 degrees to bake the coating on. There were ovens big enough to get up to the temperature we needed, but none of them had vacuum capability. Finally, working with North American Aviation, we developed a big retort (a closed laboratory vessel) that we could set the nozzle in, weld it all closed, pressure-test it, put in an inert atmosphere, and, then, put that whole business in the big oven. Remember, that nozzle was about ten feet tall and about eight feet in diameter. Facilities with the capability to heat and bake the large assembly for several hours were not readily available. Final nozzle configuration used columbium down to about the 40:1 area ratio, then titanium the rest of the way to the end. There were some quite unique issues in developing the welds between columbium and titanium. We welded them together using titanium rod. Micrographs of weld samples looked like there was no bond between the columbium and the titanium. Columbium material surface appeared just as smooth as it was before welding. But whenever we performed pull tests of joint samples, failure occurred in the weld-heat affected zone of the titanium. Whatever went on in there was strong. It worked well.

The SPS engine flew nineteen times. (See Slide 13, Appendix G) The most starts probably were performed on three of the engines, eight starts per engine, with 6,000 seconds of total time. In the overall development program, I know we tested more than 200 injectors. I think we had about 230 bomb tests. We went through 124 different injector patterns to get one that was totally compatible with the ablative material without any liners, and still approached the

> We determined that the temperatures to which we raised the columbium caused it to absorb hydrogen from what little atmosphere there was in the test cell. The hydrogen crystallized the columbium.

98 percent combustion efficiency we needed without making any grooves down the ablative chamber. All in all, the total testing during the development and qualification totaled about twenty-eight hours on the engine assembly. That's a long time for that small engine.

Just before the launch of the Apollo 8 mission, which took the first men out around the moon, Tom Paine was the acting administrator of NASA. He called George Mueller, who was the associate administrator for manned space flight. He said, "George, are you sure that engine is going to start? Find out." Mueller sent a team off to study the engine. The results of his study were the reason I knew the engine would start. He said, "There have been over 3,200 starts of that engine in development and qualification. There were only four times it didn't start, and those were all due to test stand stuff." The engine had started every time all through the development and qualification program. When I used to go to my little hometown in north Idaho, which was a long way from the space program, some of guys back there would ask me what I was working on. I told them, and they said, "What if it doesn't start?" I said, "It'll start. I'll lay it out for you. You probably start your car an average of three times a day. In a year, you've started it a thousand times. I'll bet you at least once, it doesn't go *hmmmm*, and, then, it's going to grind a little. That's one failure in a thousand." The specification I was working with was one failure in 5 million starts. So, I knew it was going to work as long as North American Aviation put that electrical cable on correctly.

Editor's Note: *The following information reflects a question-and-answer session held after Boyce's presentation.*

QUESTION: You had your fuel and your oxidizer in one valve assembly; would you do that differently now or did that work well enough?

BOYCE: It worked very well because the two independent castings had a cavity in between them. We had dual seals on each shaft where they went into the cavity/actuating area, and that area was vented. There never was a problem there. The biggest difficulty they had with respect to propellants was decontamination between the series ball valves after engine testing. There is a similar arrangement on Aerojet's Orbital Maneuvering System engine on the space shuttle. There is not a bi-propellant issue. Between the two series valves, there's always a little propellant trapped on landing, and they have to decontaminate those areas before returning to the hangar. Obviously, we didn't have that situation after an SPS engine flight. Nobody got that near to it.

QUESTION: Out of those 6,400 seconds, what percentage would you say was development versus qualification?

BOYCE: I would say probably 50 percent to 60 percent of those seconds were logged in development prior to starting qualification. There were some additional development tests in parallel with qualification testing to evaluate issues that arose during qualification. Here's a little summary of development and qualification: 216 injectors in the program, 3,200 firings, and 700 minutes of time on injectors; nozzle extensions, thirty-one units made, 1,400 firings for a total of 700 minutes. We made 274 combustion chambers, with 1,400 firings on the thrust chamber, and about 900 minutes of test time. We had twenty-seven total engine assemblies and 4,000 firings of the all-up engine assembly, which totaled about 1,700 minutes.

QUESTION: Early on in the design concept, did you have abort requirements already factored in or did those come later? That is, could you use the engine as an abort?

BOYCE: I think in one of the first Earth orbit manned launches, there was an early shutdown problem with one of the Saturn stages; they fired the SPS engine to put the Command/Service Module into orbit. It wasn't planned to be that way. During Earth orbit missions, the SPS was fired for engine operating characteristics tests, orbit corrections, and de-orbit. Remember, the original concept was that it would never fire until the astronauts were ready to come home from the moon. It did, of course, take over midcourse corrections; lunar orbit insertion and Earth return orbit insertion after they went to the LEM concept. After adopting LEM, there were no specification changes that changed the engine capability of fifty starts and 750 seconds total firing time. They did change the mission duty cycle firing schedule to match the LEM scenario. During the qualification program, an "abort" firing of 610 seconds was demonstrated. Worst-case mission requirements were thirty-six starts and 546 seconds of burn time - well within original specification.

STEVE FISHER: So, apparently it wasn't part of the requirement, but the capability was there in terms of numbers of seconds.

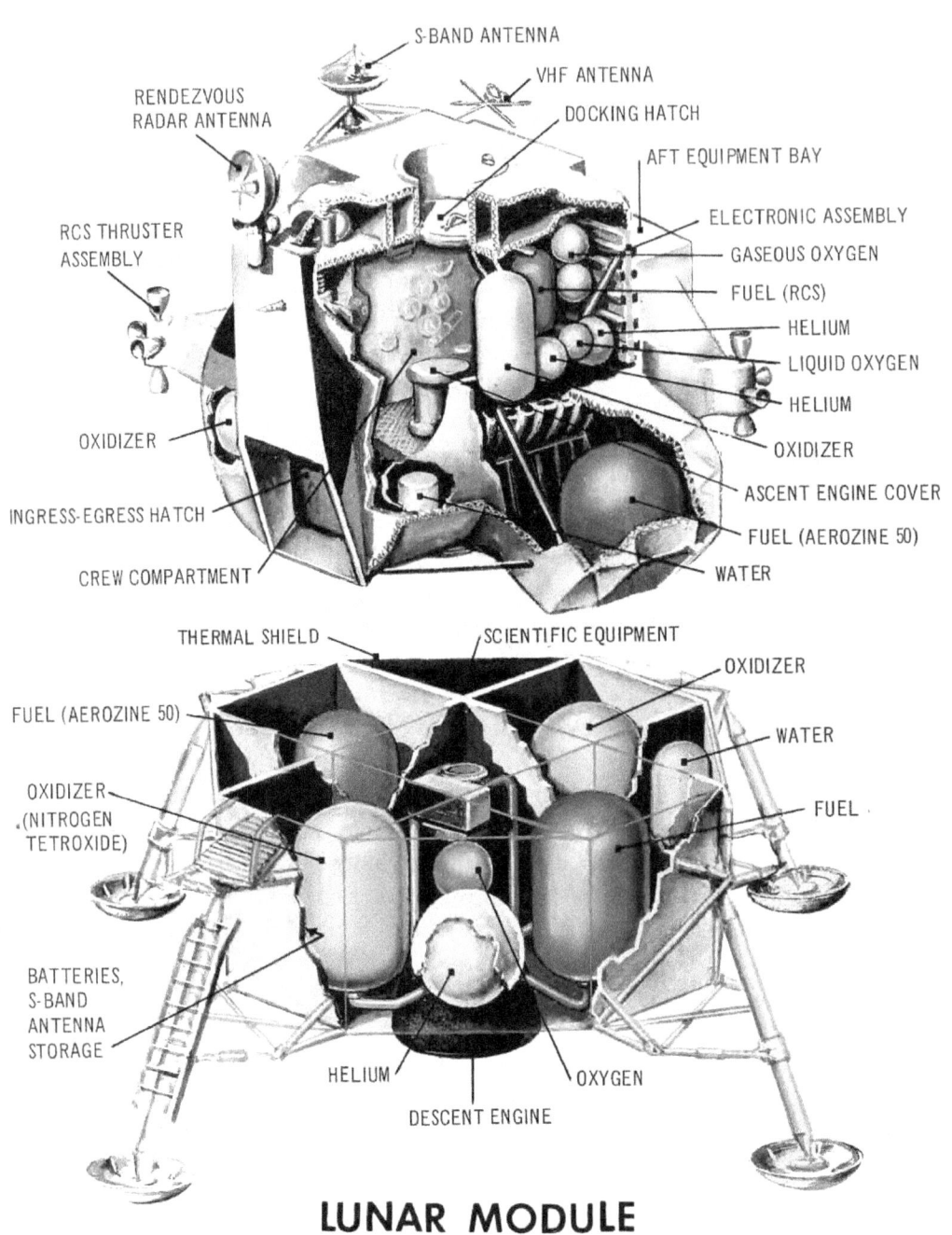

(NASA image number: MSFC-0101206)

Chapter Six

Gerard Elverum
TRW - Lunar Descent Engine

Gerard "Jerry" W. Elverum began working at California Institute of Technology's Jet Propulsion Laboratory (JPL) in 1949, where he spent ten years performing pioneering research and development on propellants and rocket propulsion. In 1959, he joined Space Technology Laboratories (later called TRW). In May 1963, Grumman and NASA selected his patented design concept for a deep-throttling liquid bi-propellant rocket engine for a backup development program for the descent engine of the Lunar Excursion Module (LEM). In December 1964, NASA committed to the Space Technology Laboratories design, and the first flight engine was delivered to Grumman in August 1966. Elverum was program director and chief engineer for the LEM descent engine throughout this time. He joined the American Rocket Society in 1951, received the American Institute of Aeronautics and Astronautics' (AIAA) James H. Wyld Propulsion Award in 1973, and was elected an AIAA Fellow in 1983. He was elected to the National Academy of Engineering in 1987. He retired as vice president and general manager of TRW's Applied Technology Division in 1990.

Chapter Six

The rocket engine companies Rocketdyne and Aerojet were the big names back in the late 1950s and early 1960s; they got the Apollo astronauts into orbit around the moon. But, how in the world did it happen that those guys were sitting in the descent module of the Apollo Program on top of an engine that was developed by a few engineers at a space engineering company called Space Technology Laboratories (STL) instead of one of the big rocket companies?

In 1960, Space Technology Laboratories was a system engineering company that was doing engineering and technical direction on the Atlas, Titan, and Minuteman programs. But, at that time, they decided they also had to get involved in space technology because they were beginning to build military spacecraft. One of the things they decided was that in order to be able to make certain maneuvers in space, and to do some of the things they wanted to do with missiles, they were going to have to come with a deep-throttling rocket engine to perform those missions. At the time, I was the head of the Advanced Propulsion Development Department at STL. They came to me and said, "We need a 20:1 throttling engine. It has to be storable. It's going to be pressure-fed. It has to be stable, and we need it in order to move into the next phase." So, I got that job. But, the story of the Lunar Excursion Module descent engine really starts back many years before that at Cal Tech's Jet Propulsion Laboratory (JPL).

At JPL, I was playing around with a pair of concentric capillary tubes. The reason I was doing that is that we had responsibility for developing the Corporal, which was the Army's intermediate, tactical, ballistic missile. At the time, they decided they had to lower the freezing point, so they went to something called stable fuming nitric acid (SFNA) from red fuming nitric acid (RFNA). The SFNA, we put a lot more NO_2, and then, we put some water in. When we did that, the Corporal engine got very rough. It had all kinds of performance problems. They turned to me and said, "You get into your laboratory and see if you can figure out what's going on." Well, I concluded that the liquid phase reactions of hypergolic propellants were really controlling what was going in the combustion of that engine, and that when we put the water in, we stopped the early reactions that led to the distribution of the propellant from the injector we were using on the Corporal. As a result of that, they no longer were mixing the way they should. They were being separated, and then, they were detonating in certain parts of the engine, and it was causing it to be rough. That meant they had to change the injector considerably and back off on how we were doing it. (See Slide 7, Appendix H)

At JPL, I was working on the liquid phase reactions with concentric tubes, and we were able to demonstrate that these early, fast reactions would occur as fast as we could mix a propellant. (See Slide 8, Appendix H) There was basically zero activation energy that was driving the system. When you have zero activation energy reactions like that, you can separate the propellants, and that was giving us problems on the Corporal. When you put some force mixing into it, you can get the reactions to go in less than a millisecond. (See Slide 6, Appendix H) But, in building that thing, it seemed like a very good idea that as long as we could get those reactions to go, we'd put that concentric tube together and see if we could make an injector out of it. If we had these early reactions going very rapidly, we didn't need distributed injection in order to be able to run the thing as a rocket engine. When we ran this with dinitrogen tetroxide (nitrogen tetroxide or nitrogen peroxide or N_2O_4) and hydrazine, we were unable to pull the inner tube back away from the face. As soon as we pulled it back in about fifty thousandths of an inch, the whole end of the tube would just detonate. That showed how fast those reactions could be because that was going off even when just a small percentage of the material was mixed. But by pushing that forward a little bit, we were able to run this engine, run it at very high combustion efficiency, and run it stably.

We were running it with N_2H_4 -hydrazine that everybody else was having terrible problems with, because it was very difficult to handle that combination from a stability point of view. When I went down to Space Technology Laboratories and they said, "Let's start a throttling rocket engine program," I determined that a fixed area injector with a separate flow control valve was one way you could throttle an engine. (See Slide 4, Appendix H) You could have a variable area injector - that was another way that you could throttle it - or you could have a variable area injector with a separate flow control valve. If you were going to vary the area of the injector, you had to be able to do that in a reasonable fashion. If you had 1,000 holes, trying to vary the area of those was going to be very difficult. But, there are problems with throttling with a fixed area injector and a separate flow control valve. If you back off on the flow rate and let the pressure drop, you need some minimum delta pressure across your injector at the low end. Then, that drives the delta pressure up so fast that you can't throttle 10:1. You just can't get there. It required tank pressures of thousands of pounds per square inch (psi), or if you start at 300 psi, you would have to have like one psi across the injector at the low end and that isn't going to work either.

The second way that you could talk about throttling an engine is with just a variable area injector. (See Slide 5, Appendix H) But, that has its own set of problems if you look at the tank pressure. Let's say we set it at 450, and we wanted a 300 psi chamber pressure. If you are going to throttle it just 10:1, this would be the chamber pressure, you get an enormous pressure drop across the injector, and the injector has to be very, very tiny. In small thrust engines, which we were working on at the time, 500-pound thrust, the dimensions around that throttling injector would be so small that we had no possibility of building it and still be in uniform flow. That was a major drawback to that type of system. But, if we go to a system that combines both a valve and a throttling injector, then we have an entirely different kind of process going on in an engine.

If you have a 300 psi chamber pressure, you are going to throttle it down to thirty psi at 10 percent thrust. You would like to be able to set the delta pressure across the oxidizer and the fuel in the injector at every part of this throttling diagram to be optimum in order to get high performance and to become stable. So, you'd like to be able to do that independently. One way to do that is to look at the pressure drop across the valve; you can go to cavitating flow. If you drop the delta pressure below the inlet to the valve below a given number, it will go into cavitation and at that point, we can set the delta pressures across the oxidizer and fuel to be any value that we want for purposes of combustion. That will totally isolate the problem of controlling the flow rate from the problems of optimizing how the injector will work. So, the one fundamental principle that came out of this was – separate the functional things whenever you can, so that you are free to optimize the function in this case, controlling the flow rates and the mixture ratio to an exact number from the functions and the processes of optimizing the injector. Then, we had what is called a cavitating venturi valve, and we designed it to be a throttling cavitating venturi. Then, we had it linked 1:1 with a single element injector in the middle of the chamber, which was optimized as far as pressure drop across the oxidizer and fuel at every flow rate to maximize the performance in the chamber. By separating these basic functions, we were able to spend our time optimizing the injector for the chamber and the mixture ratio control and the flow rate. So, over this total complex duty cycle of the Apollo descent, we knew we would have absolute control of the mixture ratio, no matter what profile they decided to run.

In a nutshell, this became the basic concept of the Lunar Excursion Module (LEM) descent engine. The injector has a sliding sleeve that controls both the oxidizer in the middle and the fuel coming down the outside. That was linked 1:1 with two cavitating venturi valves, so that for every single flow rate that was going through those cavitating venturi valves, there was an optimized delta pressure and injector configuration, totally linked together. (See Slide 9, Appendix H)

In 1962, we heard that the Apollo Program was probably going to the descent stage concept, and Rocketdyne had been given the role of building the descent engine because they were the engine guys in the United States at the time. But, we also heard that NASA was getting concerned about the basic throttling engine design and were going to go to a backup program. They wanted somebody to do a backup. Well, I took this diagram to NASA and said, "We want to bid on the Apollo engine." They said, "Who are you, and why are you in our office?" I came back with some data on this engine. It showed that we were using hydrazine and N_2O_4, which nobody else had been able to run effectively. We showed them that we had throttled it over not 10:1 but 20:1, that we'd kept a flat performance profile over the whole 20:1 throttling range, and that we never had any indication of instability or any other behavior. So, they gave us a Request for Proposal (RFP) and said, "Okay, we'll let you bid on the backup."

We had to scale that up by a factor of ten because that turned out to be the biggest thing we could put in our test stand. We only had a small test stand in the middle of Inglewood, California, at that time. (See Slide 13, Appendix H) We scaled it up to 5,000 pounds from the 500-pound engine. We built an ablative lined chamber, so that it would have to be uncooled and ran that at 5,000 pounds of thrust. Everybody was saying, "Okay, it's going to go unstable and that will put you out of business." It didn't go unstable. So, that gave NASA and the Grumman Aircraft Engineering Corporation, who was the prime contractor, a problem because now we were still in the middle of competing, and they weren't sure they wanted a little outfit like us competing. But there we were with this engine running and giving good performance and throttling 10:1 in our facility. (See Slide 13, Appendix H)

The competitors for the backup engine were Reaction Motors Inc., a liquid rocket engine company that had done the Vanguard engine, and Aerojet. One of the things they said was: "It's okay to have a 5,000-pound engine, but the LEM is going to have to be 10,000 pounds." That meant it was going to be this diameter, and if it was that diameter, it was going to go unstable and that's it. So, they said, "You've got to demonstrate with a bomb, in the right diameter chamber and show that you are going to be stable." Well, it took us a while to build that big diameter chamber, and it turned out that by the time we got it all set up on the test stand, NASA said, "We're coming out." Grumman said, "We're coming out." It was a Saturday morning, and we had never been able to fire that engine in order to see whether it was going to work. We used steam-blow in the stand in order to be able to run 10:1, and run full on the 2:1 nozzle that we were using.

We built what we called the "iron pig." (See Slides 14 and 15, Appendix H) It was seventeen inches in diameter. We ran the 5,000-pound injector into it and filled it up with bombs around in various different locations. I had to fire the first one of those tests in front of NASA and in front of Grumman, because we had just barely got it on the stand Saturday morning. Obviously, my management was not too happy with the idea of firing for the first time in front of the customer, but that was it; you either go or you didn't go. Well, we fired it, and it was completely stable. At 100 psi, we fired a bomb, went up to 210 psi. Within about ten milliseconds, we had immediately stabilized the engine at that size. Well, as a result of this test, and as a result of the fact that we were throttling 10:1 with good performance, they gave us the backup contract. It took us about a year and one-half of very intense competition with Rocketdyne, but at the end of that time, NASA came back to us and said, "You're now on for being the descent engine contractor." Well, what we had to do was promise that we would put a whole facility together at San Juan Capistrano, and we built a whole series of test stands there. (See Slides 16 and 17, Appendix H)

Chapter Six

> Armstrong finally set it down with less than a fifteen-second margin. If we had gone fifteen seconds more, he would have had to abort

For the duty cycle for the LEM, we started out at 10 percent each time. They wanted to start the duty cycle at 10 percent in order to get the vehicle stabilized. We started at 10 percent for about thirty seconds, and then kicked it up for about six or seven seconds. This was the duty cycle in order to go from the orbit they were in, into the orbit that was required to start the descent down to the surface of the moon. Then, we waited for about one hour and had the restart. We restarted again at 10 percent for about thirty seconds to stabilize all their stuff. Then, we went up to full thrust. This was the braking phase coming along on the descent. Then, we dropped down to what we called the "flare out." When we dropped from the full thrust position, we went from a calibrated full-thrust engine into the cavitating venturi region of the valves. From that point on, we could do whatever we wanted in this duty cycle. We never knew exactly what they were going to do, particularly once they got into this phase. (See Slide 18, Appendix H)

It turned out that the most critical mission we ever ran was Apollo 11. That was the most difficult landing. It exercised the engine to the absolute maximum in terms of the ablative liners capability, fuel control, and everything else. One of the things we had to do was to be able to set down on the moon and land on top of a rock. It turned out, like everything else, NASA came up with a specification for the rock this engine would have to land on and for the way the nozzle would have to crush on it. When I was sitting at Grumman during that first landing, the first thing that Neil Armstrong did when it came down and he found out they were over a boulder field, was to begin hunting around to figure out where he could find a NASA specification rock on which to land. That's because we had demonstrated the nozzle would crush okay on a rock of the specified size. At least, that's the way I felt back at Grumman. So, Armstrong kept looking and looking and looking, and I said, "Find the damn rock and set it down." We were facing problems with propellant getting ready to run out. Armstrong finally set it down with less than a fifteen-second margin. If we had gone fifteen seconds more, he would have had to abort, but Armstrong was not about to abort that flight, so he went clear to the end. We went almost to the absolute maximum of everything we could do, and I was probably the only guy in the world that knew where all these margins were and how close we were getting to them. They talked about a few guys at Houston getting blue when they finally set down. I told them, "You've got one back here at Grumman who was blue a long time ago."

The overall landing mission duty cycle involved 60 percent and 10 percent hovering modes. Talking about the descent engine, the thrust chamber design had a titanium case lined with ablative material out to a nozzle expansion ratio of 16:1. Then, we had a columbium nozzle extension from there out to 48:1. We used the Wah Chang Company to get the columbium nozzle and get the coating done. Those nozzles were very successful, but they had to be very thin. That nozzle had to be like seven-thousandths of an inch at the exit cone, and that was out about five feet in diameter. So, we stabilized it with a single flat ring in order to keep its shape. But, the reason it had to be that thin was that, otherwise, it could have kicked the whole descent stage over if it had hit some of these boulders at whatever angle. It was very successful that we were able to make that columbium nozzle work. (See Slide 19, Appendix H)

If you look at a cutaway of the final descent engine injector, you can see how the central element injector, the oxidizer, came down the middle and deflected off of the columbium tip. The fuel came down the outside of the tube, and a single moving part allowed us to change both the area of the oxidizer and the area of the fuel with a single-sleeve, driven and mechanically linked up 1:1 with the two cavitating venturi valves. We used a barrier coolant in the engine because it was ablatively lined. This engine had to run for 1,000 seconds ablatively, which is a long, long time if you are using ablative materials. We barrier cooled this just enough to keep that 1,000-second capability on the engine, and it cost us a little bit of performance. We lost about 1.5 percent of overall performance by doing that, but it allowed us to run the engine for up to 1,000 seconds. (See Slide 21, Appendix H)

When it came to the flow control valve, it was the first time I know of that anybody built a controllable cavitating venturi control valve. The reason I decided to go with a cavitating venturi on the engine was that I had used them at JPL, trying to run fluorine and hydrazine, and chlorine trifluoride and hydrazine. If you know anything about chlorine trifluoride and fluorine and hydrazine, you are never sure what's going to happen in the chamber. We would put fixed, cavitating venturis in the line in order to be sure that we were getting the right propellant flow at the right mixture ratio. We then could study what was really happening in the chamber without having what was happening in the chamber change the mixture ratio and change the flow rates all over the place. When we needed a control valve, we said there's no reason we can't take a cavitating venturi and control the area of it, which would give us whatever throttle profile we needed over the entire throttle range. That was the valve that was put on the descent engine. On the head end, there were cavitating valves on each side. The actuator was an electrical actuator done by Bendix Commercial Vehicle Systems. (See Slide 22, Appendix H)

The upper end of the valves was tied to a cross-arm, which was set on the pivot. The other end of the cross-arm goes to drive the sleeve on the center element of the injector. We went to quad ball valves on the fuel and oxidizer sides in order to assure that we had a positive chance for a shutdown or startup. The injector was spring loaded against the cross-arm, so that if we lost any electric power, the engine automatically would go to full-thrust position. That was important, because if we lost electric power on that actuator during the descent, the astronauts had to be able to go back up into orbit. The engine had to be able to go back up to full thrust where they could then turn around and go back up and, hopefully, get clear back into orbit, so they could tie up with the command module again. That was a spring-loaded system where the actuator had to drive it down to the low-thrust position. (See Slide 23, Appendix H)

The lightweight chamber liner evolved over a whole series of optimizations. We ended up with oriented silica fabric on the inside. These were oriented about sixty degrees to the flow. Then, we had a lightweight metal felt material as an insulator outside of that in the chamber, and that had a titanium structural shield around it. Then, we had a radiation exit heat shield because of the way this engine was buried down underneath the landing system where it was only the bottom part of the nozzle that really stuck out there. We had to insulate that. (See Slide 24, Appendix H)

This was a major development problem and a major lesson that we learned. When we had this thing almost finished and designed, the people who made the silica fabric in the United States decided they didn't want to have anything more to do with it. Unbeknownst to us, they got their last material from some French company, and when we got that material and ran an ablative liner, everything came to pieces. The silica, even though it was supposed to be the same spec material as we had been developing, did not, in fact, perform the way the American silica fabric worked. It took us a long time to be able to sort that out and to get back to a supplier that would provide a material that was okay. The qualification specifications on something like those composite materials were extremely sensitive, and we had to be able to watch every single detail. A little thing like somebody deciding to get their fibers from someplace else could totally change the characteristic, and that happened to us on this engine development.

The "all-up" engine had a 48:1 exit cone. The injector was in the center, and the cavitating valves sat on the outside of it. We had a square gimbal and had to gimbal the engine about six degrees. We gimbaled at the throat, not up at the head end. The gimbal worked very well. We never really had any problem with our gimbal design. (See Slide 25, Appendix H)

We did combustion stability tests; thirty-one bomb tests were conducted with five- to fifty-grain charges. We induced spikes from up to 175 percent of chamber pressure in all those tests. We had to bomb it at 10 percent, 25 percent, 50 percent and 100 percent thrust. The recovery was always less than ten milliseconds. We would give it a spike, and it would damp out immediately. We would recover in every test. At low thrust, it would take longer to recover as the pressure was much lower. But, we never sustained a high-frequency instability. In all of the engines and all of the tests throughout development and flight, we had never had the case of combustion instability in any of the central pintle injectors. As far as I know, that hasn't been the case for any other engine. In some engines, people put in baffles. They knew that we had to put them in on the Atlas engine when it went unstable on Atlas 29 and Atlas 30 in Florida and destroyed the launch stand. Those instabilities came out of the blue when people thought we had it understood. We went to baffles in the Atlas engine, but that's a damping system; it is not a dynamically stable system. It prevents the instability from growing by damping it out, and if you can damp out the disturbance, then you have to wait until it might get a new disturbance. Then, it would damp that out, but you won't racetrack it around and cut the thrust chamber off the injector. In all of these other types of engines, they either go to baffles or they go to acoustic chambers, which is another method of damping. But in the LEM engine, we have never been able to sustain an instability in that engine. (See Slides 26 and 27, Appendix H)

In my opinion, the reason is that this was what we sold to NASA and Grumman after we did the iron pig test. Why was it stable? Well, it was basically stable to the first tangential mode, because it is like trying to play a violin while your bow is bowing at the nodal point of the string. You can't make music that way. If you put the energy in at the nodal point of the tangential mode, you can't support a dynamic, unstable combustion. So, we had the central injection system operating in a stable mode, and the basic flame front was where most of the energy was being converted. If we bombed that engine, we drove that energy release zone in towards the nodal point and as soon as we drove it into the nodal point, it could not support that frequency anymore. It could not couple and drive the instability. We've done this at every size engine diameter you can imagine, up to three and one-half feet. In fact, there was one at NASA's John C. Stennis Space Center that was a 650,000-pound thrust engine operating with hydrogen and oxygen, which everybody said must go unstable. We did it at Edwards Air Force Base in California with a 250,000-pound thrust engine, where again we set off bombs, and it was stable. In every known case, both for the tangential and the radial, this injector configuration goes into dynamic stability. (See Slide 28, Appendix H)

Through March 1967, we had conducted more than 1,700 injector tests for 70,000 seconds of operation. We had twenty-six head-end assemblies overall with 55,000 seconds of operation. We did high altitude tests – twenty-seven builds, 195 different starts, and 18,000 seconds. We had a lot of experience on this hardware. But, the key to the Apollo Program success, one of the keys that I believe, is it was not a hardware-poor program. It was not, "let's just build one engine, and somehow, we'll solve all the problems as we go." It was a program that put in enough hardware that we could go to the corners of the box and find out where the failure modes were. What did it take? What were the design criteria that would, in fact, provide you

> But, the key to the Apollo Program success, one of the keys that I believe, is it was not a hardware-poor program.

with margin against the critical failure modes? You had to have enough hardware to demonstrate that was true. I think that was true of the F-1 engine program and on the service module program. We had hardware. We weren't nursing a few little pieces of hardware on some test stand some place and trying to figure out whether or not we had really uncovered all of the failure modes and demonstrated the technical design criteria necessary to assure that those modes were adequate. (See Slide 29, Appendix H)

This program had a lot of hardware in it even though it went very, very fast. The program was an extremely fast program. The Rocketdyne engine program started in February 1963. We were told that we could start our backup program in July 1963. We had to bring on a whole Capistrano[1] test facility, build all of those sites. We did our first, full throttling range test down at that site in November 1964. That was after bringing in the steam systems, and the ejectors and everything else. We finished Phase A qualification in 1966, completed Phase B qualification in 1967, and delivered our nine engines to the customer by May 1967. (See Slide 30, Appendix H)

It was a very fast program. We had a lot of hardware in it. But, in fact, they gave us the resources that were necessary in order to carry out a program as we saw we had to do. One reason for that was we had a guy named Joe Shea, who was the NASA chief engineer of the Apollo Program during most of the early critical development. Shea was one of the best system engineers I've ever come across. He would make the decisions that said, "We're not going to stick with some specification that we arbitrarily set up at the beginning of the program. I'm going to balance my specifications and my requirements, and so, if I have a limit on this component, I'm going to find out if I can change that around and take the load over here someplace else." And he did that in real time many, many times, which is one of the reasons I think we were able to get a successful program in a short amount of time, as we did with Apollo.

For any new program, one of the things you really must have is a guy who is the system engineer and is in charge of the system engineering. You have to build that discipline in such a way that you have a complete system engineering process, and you know how requirement allocation decisions are being made. The same thing is true for the Air Force now. They are relearning that same thing – you must have a system engineering process and a leader with authority. In

---

[1] San Juan Capistrano, near San Diego, is where most of TRW's propulsion test areas were located. Today, Northrop Grumman Corporation (NGC) has begun dismantling facility assets at Capistrano Test Site (CTS) after taking over the former TRW Corp. in a merger.

the end, that is what controls these kinds of things. Otherwise, you back yourself into a corner with a whole set of arbitrary things, and you don't have the money to get out from that corner. I would argue that one of the first things you must think about is the structure of your system engineering process and the authority of your chief engineer. Everything else falls out from that. In our case, Joe Shea was that guy. He did the job very well. I think he got tagged with a bum rap that was not necessarily his fault at the end of things, but during this critical part of the development, he was the guy that let it go ahead, let it go forward.

We got kind of on the wrong side with some of the guys at Houston. The astronauts didn't mind what we did, but NASA wasn't too happy. We put this billboard up on NASA Road One going into Johnson Space Center, where every time people would go in, they'd see this sign saying, "The last ten miles are on us." That's because the astronauts went into orbit for lunar descent at 50,000 feet or about 10 miles. That's where we started our descent firing. (See Slide 32, Appendix H)

Gene Cernan was the guy assigned in the astronaut corps to follow the descent engine development. He was out at our facility many times. He knew everything about the engine. He was commander of Apollo 17, the final Apollo mission. He was the last man to step off the surface of the moon. When he got back to Houston, I sent him a letter with a little drawing I had done for him. It showed two astronauts hanging onto to a descent engine as it travels to the surface of the moon. One of the astronauts is saying, "When TRW said, 'The last ten miles are on us,' I didn't realize that they were going to ride down on the engine instead of the lander." (See Slide 33, Appendix H)

Editor's Note: The following information reflects a question-and-answer session held after Elverum's presentation.

QUESTION: In looking at the statistics chart showing the number of starts, there was a ratio of about 20 percent sea level tests and 80 percent altitude tests in the descent engine program. Maybe that ratio was comparable for the other engine, the Aerojet engine. But, given that, let's say that thrust is an important parameter that you had to measure, especially with the throttling 10:1. We often say, "Test like you fly, and fly like you test." In terms of this engine, how did you know that the facility thrust measurement and the facility altitude simulation were sufficient to replicate or represent flight conditions?

ELVERUM: As we went through the program, what we determined, and what we all agreed on, was that the thrust coefficient (C_f) of the nozzle, after you get past a certain point, is really an engineering parameter. It's not a fundamental parameter that is going to be highly variable. Once we knew what the contour of the nozzle was, and once we knew what its characteristic was out to 2:1, we could calculate what the 48:1 thrust coefficient was going to be. In every case that we made a test, the calculation was precise. We weren't looking for a problem out at 48:1. Once we crushed the nozzle and said, "Yeah, we can land on the boulder," and once we had the thermal profile of that columbium nozzle, we did not require a lot of effort there. The real characterization was done in throttling over the 10:1 with the injector and controlling the mixture ratio on that – the whole head-end assembly – out to 2:1. I think everybody at NASA and Grumman agreed that flying like you test is great, particularly if you are using an aircraft engine. But, in this case, the thrust coefficient of the nozzle was not an issue.

QUESTION: How many engines did you test in the altitude chamber?

ELVERUM: I can't remember. We had twenty-seven complete 48:1 engine builds that were acceptance tested and tested in various types of throttling programs, which is a lot of hardware. I've been following the SpaceX program, for example, as part of the role that I do for the U.S. Department of Defense. There, they just have a few engines. And, a whole bunch of other commercial rocket companies who want to get in the business talk about, "Well, if I could get a little more money, I would be able to build two engines." They think that's going to demonstrate that we ought to put a 500 million-dollar payload on top of a space vehicle that's had two engines tested and only tested to nominal conditions. They never had a chance to burn out three of the engines and go way off nominal.

It takes a lot of hardware to be able to demonstrate that you have understood the criteria that established the risk of an engine. The risks of an engine means you have to define the failure modes and you have to do some real "honest-to-God" risk assessment. I mean *quantitative* risk assessment, not just qualitative. By quantitative, I mean you assess what the profile of the risk is, then you test that profile and see if you can collapse the uncertainty of that risk. That takes a lot of hardware. It's something that has to be put into a new program. We did all kinds of stuff like that, but it is very expensive, very expensive.

QUESTION: Do you have any comments on how the descent engine was used in Apollo 13?

ELVERUM: In Apollo 13, it was very strange. I've always had a kind of an eerie feeling about the whole program because when I was at JPL as a young engineer working with hypergolic propellants and cavitating venturis, I had no feeling or concept of an Apollo Program. Apollo

5 was the first time that we ran the descent engine in space, and it was hooked up in space after separating from the S-II.

We had the tandem configuration of the service module, the command module, and the LEM sitting out there, and we were to fire the LEM. On Apollo 5, we were firing the LEM to show how it would work. There was a problem. I can't remember where the problem was, but something caused a problem before that engine had finished its burn. It was not in the engine, but there was some other problem, and NASA made a controlled shutdown. Then, they came to us and asked, "Hey, we're up there. We want to finish this test program. Is it okay if we restart that engine again in space with this tandem configuration?" We said, "As long as it has been more than forty minutes since you shut down, our analysis says that you will be okay in terms of the thermal characteristics of the inside of that chamber." They restarted it and pushed that system around in orbit on Apollo 5.

It turned out, that when it came to Apollo 13, we went back into the record, and said, "Hey, we have pushed this system around up there on Apollo 5, and we have also restarted this tandem configuration." The requirements on Apollo 13 were to put it back into play. The spacecraft was out of free return to the earth at the time of the accident. It would not have come back. NASA said, "Okay, we'll use the descent engine to put the spacecraft in a free trajectory; it will go around the moon and be on free trajectory back to Earth." Then, as it came around the far side of the moon, the guys found out that they had an oxygen problem. As you remember, things were getting pretty bad in there. They said, "We've got to get it back as fast as we can. Is it okay if we re-fire the engine? Now, we're in a free trajectory, so we want to put as much delta-v (or change in velocity) in as we can. Can we re-fire right now?" We said, "Yes, the data says it has been this period of time." We could re-fire the engine, run the rest of the duty cycle up as far as we needed while preserving enough fluids to make the final correction as the spacecraft got near Earth, and restart the engine. It was pretty fortuitous that we could give them those answers.

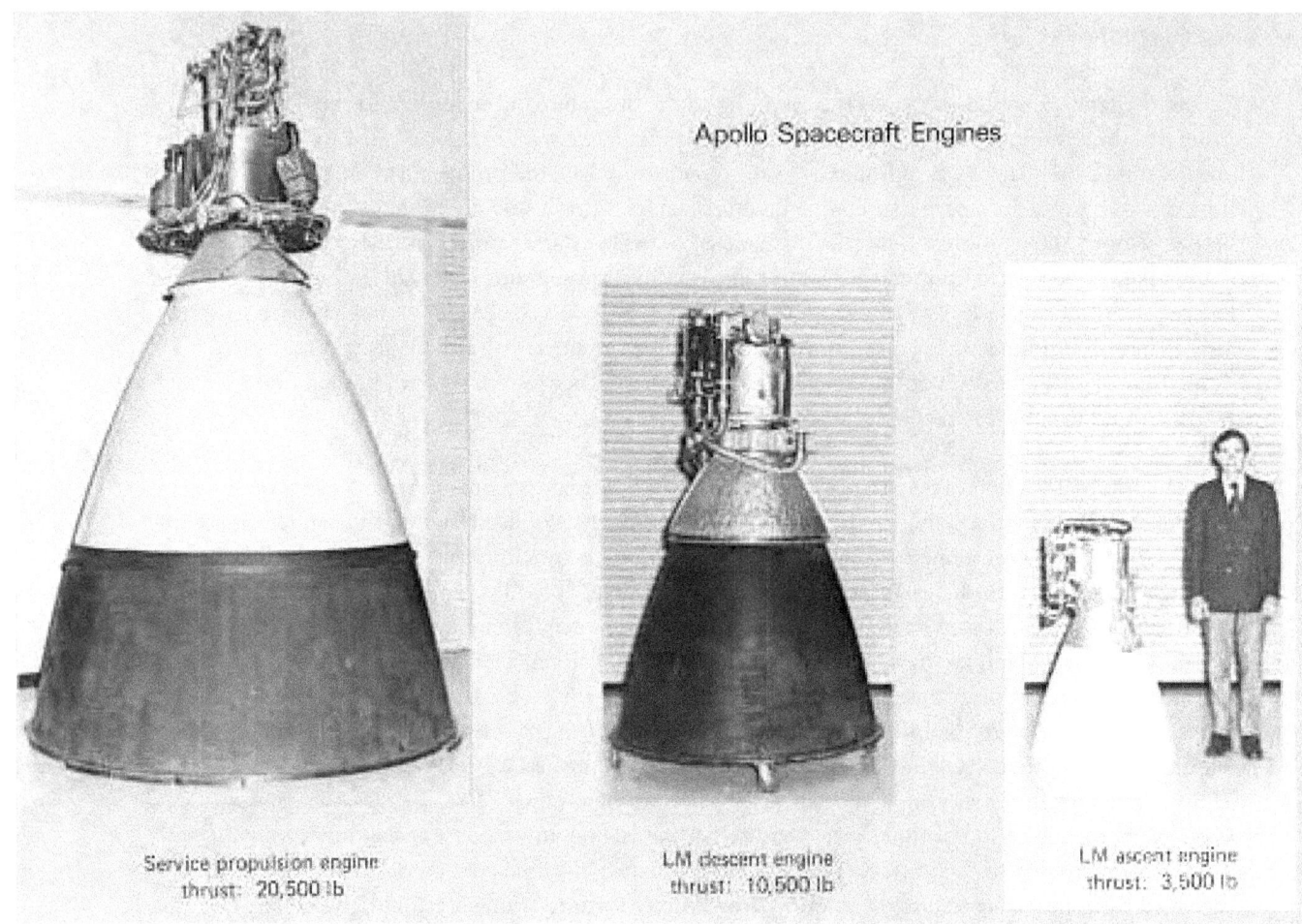

Similar in shape but not size were the three big engines aboard Apollo spacecraft. Two of them had no backup, so they were designed to be the most reliable engines ever built. If the service-propulsion engine failed in lunar orbit, three astronauts would be unable to return home; if the ascent engine failed on the Moon, it would leave two explorers stranded. (A descent-engine failure would not be as critical, because the ascent engine might be used to save the crew members.) (Source: http://history.nasa.gov/SP-350 - Apollo Expeditions to the Moon", published 1975)

Chapter Seven

Tim Harmon
Rocketdyne – Lunar Ascent Engine[1]

The ascent engine was the last one from the moon, and I want to focus on the idea of redundancy and teams in regard to the engine. By teams, I mean teamwork – not just within Rocketdyne. It was teamwork within Rocketdyne; it was teamwork within Grumman; it was teamwork within NASA. These were all important elements leading to the successful development of the lunar excursion module (LEM) engine. Communication, rapid response, and cooperation were all important. Another aspect that went into the development of the ascent engine was the integration of technology and of lessons learned. We pushed all the above, plus technology and lessons learned, into a program, and that led to a successful result. One of the things that I like to think about – again in retrospect – is how it is very "in" now to have integrated product and process teams. These are buzzwords for teamwork in all program phases. That's where you combine a lot of groups into a single organization to get a job done. The ascent engine program epitomized that kind of integration and focus, and because this was the mid- to late-1960s; this was new ground for Rocketdyne, Grumman, and NASA.

Redundancy was really a major hallmark of the Apollo Program. Everything was redundant. Once you got the rocket going, you could even lose one of the big F-1 engines, and it would still make it to orbit. And once the first stage separated from the rest of the vehicle, the second stage could do without an engine and still make a mission. This redundancy was demonstrated when an early Apollo launch shut down a J-2 second-stage engine. Actually, they shut down two J-2 engines on that flight. Even the third stage, with its single J-2 engine, was backed up because the first two stages could toss it into a recoverable orbit. If the third stage didn't work, you were circling the earth, and you had time to recover the command module and crew. Remember how on the Apollo 13 flight, there was sufficient system redundancy even when we lost the service module. That was a magnificent effort. TRW Inc. really ought to be proud of their engine for that. (See Slide 2, Appendix I)

We had planned for redundancy; we had landed on the moon. However, weight restrictions in the architecture said, "You can't have redundancy for ascent from the moon. You've got one engine. It's got to work. There is no second chance. If that ascent engine doesn't work, you're stuck there." It would not have looked good for NASA. It wouldn't have looked good for the country. There was a letter written that President Richard Nixon would read if the astronauts got stuck on the moon, expressing how sorry we were and so forth. It was a scary letter, really. The ascent engine was an engine that had to work. (See Slide 3, Appendix I)

---

[1] Please see Section 4 for Mr. Harmon's biographical information.

Chapter Seven

> Reliability was the name of the game; keep it simple. That was the architecture and the plan.

The lunar excursion module vehicle was an ungainly, lightweight vehicle. The vehicle walls were really thin pieces of metal. The ascent engine was buried inside. All you could see was a little bit of the nozzle extending from the bottom of the vehicle. (See Slide 3, Appendix I) The LEM vehicle ascent stage height was twelve feet, and the diameter was thirty-one feet. The LEM weighed about 11,000 pounds. Now, the ascent engine was fifty-one inches high; it wasn't a very big thing. It produced 3,500 pounds of force (lbf) thrust. It was a very "cozy" environment, this LEM vehicle. By cozy, I mean it had to be lightweight. By cozy, I mean they couldn't even install seats due to the weight penalty. The astronauts stood! They had parachute-like harnesses to provide stability. The astronauts stood and looked out "tilted" windows to see the landing area. The descent engine had a backup, the ascent engine, and the reaction control engines on the vehicle (there were four sets of four) were redundant. Again, the situation was: we did not have a backup for the ascent engine. When you don't have a backup, you make the thing simple. NASA was smart in this. They said, "All we want you to do is one on/off switch…. You don't have to throttle – constant thrust is okay. We are going to fix the mount and we are not going to gimbal this engine. We are going to have redundant valves and hypergolic ignition. Get the two propellants together, it lights." This is the plan for a simple, reliable engine. That was the approach, and I think it was a sensible idea. (See Slide 4, Appendix I)

The specifications for the engine were: pressure-fed, 3,500 pounds of force vacuum, a 310 specific impulse (Isp), and hypergolic propellants. One possible hypergolic fuel, hydrazine, freezes like water at two degrees Celsius. This was unacceptable, so unsymmetrical dimethylhydrazine – I don't know who comes up with these names – suppressed the freezing temperature. Reliability was the name of the game; keep it simple. That was the architecture and the plan. (See Slide 5, Appendix I)

It's important to remember Rocketdyne did not start out with this contract. This was the Bell Aerospace contract, which began in July 1963. Bell had challenges. Every development program does. Their biggest issues were thrust chamber erosion and combustion instability. These continued in the program all the way up to 1967. If you followed the descent engine program, you knew it was through qualification testing (also known as "qual") in 1967. In 1967, Bell hadn't even started qualification testing. So, suddenly NASA realized they could get to the moon, but they might not be able to get off the moon – not a good situation. NASA awarded a backup program to Rocketdyne in August 1967. We were successful with the main issues. By 1968, we had won the complete contract and were off and running. A calendar was established to set the major development schedule milestones. (See Slide 7, Appendix I) The Bell program was given to NASA in July 1963. The Agena engine was Bell's basis for the ascent engine. It was a reasonable upper stage engine and had several successful satellite launches. I think it was a 16,000-pound thrust engine. In late 1964, I believe, NASA said, "We want to do stability testing. We want the engine duration to be longer." So, the specifications changed a little bit, and Bell began stability testing. The original injector was unbaffled. As soon as they bombed it to test stability, they had instability. They put a baffled injector in to improve stability, but were not successful in eliminating the instability problem. That created a situation in which Rocketdyne became a backup program. (See Slides 6, 7 and 8, Appendix I)

Because I did not work for Bell Aerospace, I can only provide my impressions of the ascent engine development problems based on NASA's historical documents. One of Bell's development issues was their program had limited test hardware early in the development cycle. When Bell had instabilities, there were manufacturing issues that could be blamed. You also get the impression there was some denial that there was a problem. When engine combustion went rough, they were never sure of its cause. "Was it manufacturing; did we do something wrong there? If we had an ablative wall thrust chamber, instead of steel battleship-type thrust chamber, maybe the ablative would change the acoustics, and it might absorb it." Instabilities tend to do a lot of hardware damage. I think their rough combustion cutoff was a little slower than Rocketdyne's, or what would be typically used, so they sustained a fair amount of hardware damage by the time they shut the engine down. It didn't quite tell them what the issues were. All those variables added up to mask the root cause of their problem.

Rocketdyne had heads-up information the ascent stage was in trouble. There's a story I can't deny or confirm: that Rocketdyne marketing people at NASA's Johnson Space Center in Texas had the ability to read documents upside down on a person's desk. Doing so, they realized there were problems with the Bell ascent engine. It did not take a rocket scientist to realize that problem, because in 1967, Bell hadn't started qualification testing. Rocketdyne thought, "There's something wrong here. Maybe this is an opportunity." It did prove to be an opportunity. NASA recognized the issue and said, "We need a backup ascent engine system, just in case." They put out a Request for Proposal in June 1967 for a backup injector system that would address the stability issue. Three companies were looked at and down-selected. TRW might have been one; but I'm not positive about that. The others were Aerojet and Rocketdyne. Rocketdyne's approach was to put everything we could think of into stability. We decided we had better design in baffles. We incorporated the best manufacturing technology that was available. We went with electron beam (EB) welding. We incorporated electron discharge machining (EDM) drilling, which is a real nice, advanced, repeatable system for drilling orifices. Then, we put in acoustic cavities.

Acoustic cavities were another stability aid. Rocketdyne had, in 1965 or 1966, an independent research and development program on a small rocket engine. It used a beryllium thrust chamber with a stainless steel injector. Beryllium is a brittle material. If the injector gets warm, it will expand, and the brittle beryllium material will break. To make up for that, engineers put a gap between the injector and the thrust chamber. We tested this in a pulse mode multiple times.

This was a small engine, just a little pulse mode engine. It turned out this little beryllium engine with a stainless steel injector occasionally had low performance pulses; i.e., the pulses wouldn't go quite as high on the Datelink Independent Gateway Retrofit (DIGR). Mostly, pulses were to be a certain height. Occasionally, 3 percent of the time it turns out, pulses wouldn't go quite as high. We didn't understand that. We didn't know what was going on there. There was no high-frequency instrumentation. It was just that we noticed this data anomaly. The hardware didn't break, and there wasn't a problem with the hardware. We thought, "It is working and it isn't broke; let's let it go." But management said, "We have to understand the problem." During the analysis, we thought, "Well, maybe it's that gap between the injector and the beryllium chambers, and propellants are sneaking up there and detonating and causing some sort of instability." The question then became: "Make the gap smaller or make it larger?" We chose to make the gap larger, and cut the diameter of the injector down. The instability disappeared. They were all now the same height. Engineering said, "Aha! We've got the problem solved." Management said, "We don't think you do because you don't know what you did." We had to analyze it.

We then decided maybe it was a Helmholtz-type resonator disrupting the instability. Most of our engines in 1965 and 1966 were in production. We were not going to add acoustic cavities, which was a precursor of this type of stability aid. That idea went on the shelf. Then, along came this opportunity. We had come up with this injector and put a baffle in it. The president of the company, Sam Hoffman, said, "Let's put those acoustic cavities in there." I'm not sure we had all the analytic background, but that seemed to work on this little internal research and development (IR&D) engine. Hoffman said, "Put it in there." Of course, we already had our design, and we had to do some redesign to get the acoustics cavities in there, but it was a brilliant idea, and it helped sell the program to NASA as the backup program for the Bell problem.

People talk about the competition at that time. It's true we were in a space race, and we were doing pretty well maybe by 1966 or 1967. But NASA said, "Rocketdyne, you are still in competition with Bell." To set the stage, Bell was conducting their development program in parallel after we won the backup contract, which put us in competition. We had the competition of the space race and the competition with the schedule. Near the end of 1967, we hadn't even started qualification testing on the ascent stage. Rocketdyne had to respond to this competition. They recognized right off that this program had to be an integrated effort, and put together a team consisting of engineering, development engineering, design, and other personnel. It included inspection and quality teams. It included NASA and Grumman. It included a member of the test stand team. The test organization was represented there. They were all in one place, on one team.

Every morning at eight o'clock, we had a standup meeting. It was truly a standup meeting. The offices in those days were fairly small. At eight o'clock, people piled in there. Every morning, it was like trying to see how many people you could stuff in a Volkswagen. For fifteen minutes, they discussed the results of yesterday, what we would do today, and marched on. It was a twenty-four/seven type of operation. Rocketdyne really went on this twenty-four/seven operation with a vengeance. Because of schedule constraints and schedule pressure, the team pushed down the point at which decisions were made. This was 1967; I had only been with the company four years, a fairly junior engineer. I was in charge of stability testing, which was run in two shifts. The first shift and second shift were stability testing. The third shift cleaned up the mess we made in the first and second shifts; then, it started all over the next day. I was the development engineer, and my job was to get stability tests done as rapidly as possible. They said to me, "If we can gain schedule, you can work the crew weekends – Saturday and Sunday." That part of my charter was a lot of responsibility for a guy of four years' experience because the test and operation staff was thirty people a shift. I was a junior engineer able to say, "We're going to spend some money this weekend to gain schedule." We were really pushing. We had the long pole in the tent for 1967. I don't mean that I was particularly special; that was the way it was for everybody on the team.

This was one of the most fun and fulfilling programs I ever worked on; it was a very intense program, but let me describe what Rocketdyne was able to accomplish. The contract proposal was submitted in June 1967. It was awarded as a backup in July 1967. We got a go-ahead in August. However, once we submitted the proposal in July, we recognized the need for speed. We started assuming we would win the contract, and we started building hardware. It's a phenomenal thought, but we were able to design and get ready to go in two months.

We conducted our first altitude test, then our first full duration test two months later. We completed our preliminary design review in November. We completed our critical design review in December, the first engine system full duration altitude test in January. The term "design feasibility tests" meant we knew enough of what we were doing regarding stability, to finish proving that in March. Our development program milestones give some sense of what we were able to accomplish, and we were not hardware poor. (See Slide 8, Appendix I) We were able to get a lot done in a very short period of time. (See Slide 9, Appendix I)

For the design feasibility program, we had twelve injectors. We accomplished 872 tests, including 302 combustion stability tests. The combustion stability test used a little bomb. It was more like a blasting cap or a firecracker, but it was a little bomb. We built a battleship, just a plain steel thrust chamber. We could run this chamber about ten seconds before it got too hot. I could get five bomb tests off during a ten-second run. With every hot-fire, we could get five stability tests accomplished. NASA and Grumman were coming in to review the program one Monday. We were going to work through the weekend to get as many bomb tests off as we could. We wanted to have 100 bomb tests completed. We got ninety-nine tests done, and then, it went unstable. I couldn't believe it. I was crushed. It turned out the instability was caused by the bomb shrapnel damaging the injector. In post-test, we disassembled the injector and recognized we had damaged the injector with shrapnel from the multiple bomb detonations. The wiring to the blasting cap was going backwards, hitting the injector face, changing the injector orifices' shapes, and denting the injector. After that, we would take the injector apart after each test, and if we saw any damage to any orifice we would hand drill it back out. That eliminated the instability problem. But, it made for a very interesting NASA meeting that Monday because the injector went unstable Saturday night about eight o'clock, and we only had Sunday to figure out what caused this instability and whether we understood it. (See Slides 10 and 11, Appendix I)

To give an idea of the twenty-four/seven intensity of the schedule, Rocketdyne decided to appoint only one development engineer for stability, and it was me. I don't know why. But, to accomplish testing like that, I was to be available whenever the engine ran. In those days, I wore contact lenses and really couldn't wear contact lenses much more than ten hours. I just couldn't see beyond that. I had to do something to be able to see twenty-four hours with contact lenses. So, I would wear one contact lens at a time. I would look this way, or I would look that way. By three or four o'clock in the morning I knew it was time to change because my eyes were really getting tired. I was so tired, I managed to stick both contact lenses in one eye. It took a while to straighten myself out, but I did get the contacts sorted out and completed the test program. That shows you some of the dedication of the team. We really worked at it.

Another program issue was combustion chamber erosion. There were a couple of others as well. One was baffle weld cracks. Last, we had to make the engine lighter. Within nine months of becoming a parallel effort with Bell, Rocketdyne had gotten to the point of showing NASA that we knew what we were doing, and we became the single contractor for the ascent engine. Within nine months, we had caught up with Bell, who was probably four years of testing ahead of us, and we became the prime contractor for the ascent engine. That was quite an accomplishment.

Finally, with combustion instability, we did one test where we verified that the acoustic cavities were doing their job. On one test, we blocked the acoustic cavities, and sure enough, on the first bomb the injector went unstable. The acoustic cavities were validated with that test.

The chamber erosion issue was very difficult to solve due to the need of getting uniform flow from the injector. One way to protect ablative material is to put film cooling material down the walls of the thrust chamber. But the engine loses performance when you do that. There is always a trade off. We wanted to protect the thrust chamber, but we couldn't do too much because the specific impulse performance goes down. After we had several tests that showed erosion, we discovered it was always occurring in the same general area. With the conventional wisdom of putting film cooling all around, we would end up with unacceptable performance. My manager, Mike Yost, said, "Well, you know, we don't need to put film cooling all the way around, we just need to put it where it is eroding." That really sounds like an easy solution now, but it wasn't conventional at the time. In those days, you put film cooling in a nice symmetrical pattern. But Yost said, "Let's just do it where we need it." That eliminated most of the film cooling performance loss. This was a good example of thinking out of the box.

I mentioned that we had hard starts. That happens when the fuel gets to the combustion chamber before the oxidizer. When the oxidizer hits, it goes "bang," which is not a good situation. Our solution was to make the fuel pipe large and add an extra little volume at the end of the fuel pipe, so that the fuel had to prime a large duct. When we did that, the oxidizer reached the combustion chamber first, before the fuel. The design looks pretty unconventional and is not real elegant. (See Slide 12, Appendix I)

Testing was key to demonstrating high engine reliability. We had test facilities everywhere. There were two test sites at Santa Susana in California. We used White Sands, New Mexico; Reno, Nevada; and even Bell Aircraft test facilities. We had to take our engine to Bell and run it there. Bell was, of course, extremely interested in our design. We were still in a competitive situation. After they installed it, they thought it would be a good idea to X-ray the engine. We said, "No way." They did focus on our unconventional inlet duct, thinking it was a stability aid. It was only a priming aid to keep the hard starts from occurring. If they thought it was the solution to stability, it wasn't – perhaps it led them down an incorrect thinking path. It was an accidental thing, not done on purpose on our part. But it did cause Bell a lot of consternation. We ran successfully in their facility, partly to verify that our performance was equal to theirs. It was.

The LEM engine was on the critical path. It really was a twenty-four/seven operation. We had this team that included everybody. It really was an integrated effort. *Another milestone or key part of the effort was communication. It was rapid. It was effective.* Decisions were made in timely order for anything. Decisions were down to the lowest person. It was interesting that we had this close working relationship with Grumman Aircraft, as they had chosen Bell, but they recognized the issue and so did Bell Aircraft. I should also say that Bell was very helpful in the program. We had the injector. We assembled the engine. But the valves and the thrust chamber were Bell Aircraft's. Everybody worked in concert, because the goal was to get to the moon and to get back. (See Slide 13, Appendix I)

> Testing was key to demonstrating high engine reliability.

When trying to understand how fast we were able to respond, it's important to remember we were not hardware poor. We essentially got going within a month of the contract award. Our feasibility testing was completed. Then, the solution was demonstrated literally in eight months. This was a prototype integrated product and process development (IPPD) team. It was a totally committed, very goal-focused team, in which all the disciplines were involved. I think Rocketdyne really showed what they could do in a critical situation. I think we did put the lessons learned in this engine. We implemented them in an effective manner, and we integrated technologies within that environment. Once we got the engine working, NASA asked us to reduce its weight. This engine only weighed 171 pounds, but still, the weight had to be reduced. The reason was that NASA wanted to bring home more rocks from the moon. If the engine weighed less, they could do that. I was put in charge, and we took a 171-pound engine and reduced its weight by thirty pounds. That meant an additional thirty pounds of moon rocks could come back to Earth. That was a significant amount of improvement. That was a 15 percent weight reduction on a pretty well-developed engine. (See Slide 14, Appendix I)

Editor's Note: *The following information reflects a question-and-answer session held after Harmon's presentation.*

QUESTION: One of the things that we run into with the "can-do" attitude that you showed on your last slide is you don't want to sacrifice quality, and that's always the issue. How do you balance that? How did you balance that back then to make schedule and still maintain quality? What were you doing? What advice would you give?

HARMON: One way was that quality was part of the team. When you put manufacturing, quality, and engineering personnel together, quality can look at your design work and say, "Man, that is really hard to see whether that is going to be a good weld or not." They are right in tune. I was talking recently with some gentlemen who had testing personnel as part of their team. They were there every day. They were there in the standup meeting. When an engineer said, "I need 100 strain gages," the test guy said, "I got thirty. You are going to have to make do with thirty, or it is going to cost you 'X' bucks to go to sixty and it's going to take 'X' months to get it available for you." We had this immediate feedback of what we were asking for, whether it was doable, whether quality staffers could do it, and that was the virtue of that team. The quality check certainly was built into the design as best we could do.

There was one instance where we found weld cracks on this system. There is an art to welding. This piece was machine-welded, but there's still an art to it in the setup. What was the problem? Well, it turned out, we were working two shifts, and the machinist on the night shift didn't quite make the setup right. It was a difficult weld. It was difficult enough that we would sometimes overstress that weld if the setup was not perfect. It was a weld penetration issue. We solved the weld problem by removing the second shift welder from the task.

I got to do the inspections on that weld, to see whether I felt the penetration was correct. We actually had one of our LEM engines in a "bird" (referring to an Apollo vehicle on a Kennedy Space Center launch pad) at Florida. Since it was not a big engine, it was hard to look from the outside and see how good the weld penetration was. It was supposed to be a one-day trip.

The welding, though, was interesting because of how we were able to solve that issue and figure it out quickly and verify the welds on the LEM were fine – the same way I verified the quality of that paper swimsuit.

Rocketdyne met the challenges. The Lunar Excursion Module Ascent Engine was a prototype IPPD program, but it was probably the best yet for integrating lessons learned and technology. It was one of the best programs I ever worked on. I think almost every team member who worked on the LEM looks back on it, and is very pleased with the results.

Appendix A

Event Program

National Aeronautics and Space Administration

"On the Shoulders of Giants"

John C. Stennis Space Center
StenniSphere Auditorium

Tuesday, April 25, 2006
8:45 a.m. – 3:30 p.m.

—Speakers' Biographies—

Robert Biggs ("Bob") has worked for 47 years at Rocketdyne, and spent 9 years working as lead development engineer and development project engineer on the F-1 Engine Program. He spent several months on the cancelled Navajo project, 3 years as lead engineer in Jupiter performance analysis and a year as manager of the Dynamic Analysis Laboratory. Mr. Biggs also has worked for 34 years on the space shuttle main engine, serving as development manager and chief project engineer.

Paul Coffman earned experience during many Apollo-era assignments. He served as lead engineer on the J-2 Thrust Chamber Assembly Development, supervisor of engineering test for J-2 components and engines, and manager of J-2 Engine Development and Flight Support. Following the Apollo-era, some of Mr. Coffman's program management and marketing assignments have included serving as director of Controls Engineering, program manager of the Compact Steam Generator, engineering coordination for the space shuttle main engine, and project engineer on the Gas Dynamic Laser.

Carl Stechman began his career at Rocketdyne in 1958. In 1959, he was hired at Marquardt as a heat transfer engineer (Sacamjet and LACE Cycles), and then worked from 1964-1967 in the Apollo Program as manager for development engineering. Mr. Stechman next served as project engineer for the Space Shuttle Primary and Vernier Thruster, and then as program manager for the High Performance Iridium Rhenium Engine. He has now worked for 45 years for the company, now known as Aerojet. Mr. Stechman earned his masters degree in chemical engineering from the University of California Berkeley, and his masters in engineering from California State University-Northridge.

G. R. Pfeifer ("Jerry") began work at the Marquardt Co in 1961, performing early work to develop the parametric rules used for Marquardt rocket design and, with a small group, built their first hypergolic rocket test facilities and rockets. When the Apollo RCS program came to Marquardt, Mr. Pfeifer helped shepherd that program through its Qual program, which led to many other successful NASA and other space propulsion programs, including its most well known – the space shuttle. Aerojet now provides the same shuttle attitude control engines to NASA. Mr. Pfeifer went on to work at Rocketdyne to help develop their THAAD attitude control engines, and is now working on the Airborne Laser Program at The Boeing Co.

Gerard W. Elverum ("Jerry") began working at Cal-Tech's Jet Propulsion Laboratory in 1949, where for 10 years he performed pioneering research and development on propellants and rocket propulsion. In 1959 he joined Space Technology Laboratories (later called TRW). In May 1963, Grumman and NASA selected his patented design concept for a deep-throttling liquid bi-propellant rocket engine for a backup development program for the Descent Engine of the LEM. In December 1964, NASA committed to the STL design, and the first flight engine was delivered to Grumman in August 1966. Mr. Elverum was Program Director and Chief Engineer for the LMDE throughout this time. He joined the American Rocket Society in 1951, received the AIAA's James H. Wyld Propulsion Award in 1973, and was elected an AIAA Fellow in 1983. Mr. Elverum was elected to the National Academy of Engineering in 1987. He retired as Vice President and General Manager of TRW's Applied Technology Division in 1990.

Clay Boyce left the U.S. Air Force in 1955, and joined Aerojet at its Azusa, Calif. facility. After Sputnik, he joined the Thor/Able 2nd Stage Program, an adaptation of the Vanguard 2nd Stage. Subsequently, he participated in the conceptual design and development of the AbleStar Upper Stage. Variations of this stage are still flying. In 1960, he was assigned to the Win Apollo team and then became engineering manager of Apollo SPS Engine development. In 1969, he was assigned to Johnson Space Center for technical support of the SPS Engine. Later assignments included the Space Shuttle OMS Engine, Japanese N-2 Upper Stage and the National Aerospace Plane. Mr. Boyce retired from Aerojet in 1991, and has provided consulting services to Aerojet since that time.

Tim Harmon has more than 41 years of experience with Boeing, Rocketdyne. He retired as the Chief Systems Engineer on the MS-XX Program, a cryogenic upper stage propulsion system. Mr. Harmon was also involved in 10 Rocketdyne engine development campaigns, ranging from a large Aerospike engine to small attitude control engines. He has first-hand experience on the development of the Apollo Program Lunar Module Ascent and Command Module Attitude Control Engines. Mr. Harmon earned a bachelor of science degree in aeronautical engineering from Purdue University, a masters of science in mechanical engineering from University of Southern California and a masters of business administration from Pepperdine University.

Moderator: Steve Fisher has spent 34 years at Rocketdyne in the design, development, testing and evaluation of liquid rocket engines, and is now a Technical Fellow of Pratt & Whitney Rocketdyne. He initially joined North American Aviation of El Segundo to work on the original B-1 Bomber, and later transferred to the original Rocketdyne Division of the company. During his tenure, he was integrally involved in rocketry at the Coca 1 and Coca 4 test facilities of Rocketdyne, and has further contributed to a wide variety of propulsion activities including R&D projects with storable and cryogenic propellants, as well as flight engine projects such as the RS-68 engine design and development. Mr. Fisher presently serves as the Senior Principal Engineer for mechanical design at Pratt & Whitney Rocketdyne.

Appendix A

40th Anniversary of the First Rocket Engine Test at Stennis Space Center
April 23, 1966

- On May 25, 1961, President John F. Kennedy made the historic announcement that the United States would put humans on the moon by the end of the decade. The plan called for a place to test the powerful rocket engines that would transport them.

- The site's water access was essential for transporting components, propellants and the large rocket stages that came from Michoud Assembly Facility in New Orleans. Its vast land – 13,500 acres – is protected by a 125,000-acre acoustical buffer zone, and today is considered a national asset.

- Mississippi's influential Sen. John C. Stennis (Aug. 3, 1901-April 23, 1995), for whom Stennis Space Center is named, promoted his home state as the place to build the rocket engine test facility.

- Construction began in May 1963. The site – under its original name, the Mississippi Test Facility – was fully operational for rocket engine testing in less than three years.

- Stennis Space Center's primary Apollo mission was testing the first and second stages of the Saturn V rocket.

- The S-II cluster of five engines that powered the second stage of the Saturn V craft were incredibly powerful – 21 million horsepower

- Just three years after the first engine test at Stennis Space Center, on July 20, 1969, astronauts Neil Armstrong and Buzz Aldrin walked on the lunar surface

- Stennis Space Center continued to test the first- and second-stage Saturn V for the Apollo program until the early 1970s, the dawn of the space shuttle program. Then, Stennis Space Center's rocket engine test stands were modified to test-fire and prove flight-worthy all space shuttle main engines. The first space shuttle main engine hot-fire test was performed June 24, 1975.

–PROGRAM–

Welcome
Dr. Richard Gilbrech, Director, Stennis Space Center

Introductory remarks
Steve Fisher, Technical Fellow, Rocketdyne
Shamim Rahman, NASA Stennis Space Center

F-1 Booster Engine
Bob Biggs, Rocketdyne

J-2 Saturn V Second and Third Stage Engine
Paul Coffman, Rocketdyne (retired)

■ 10:15 a.m. – Break
Program resumes at 10:35 a.m.

Attitude Control Engines
Carl Stechman, Marquardt (Aerojet Redmond Operations)
Presenter: G. R. Pfeifer

SE-7 (S-IVB Stage APS Ullage Control) & SE-8 (Apollo Command Module RSC) Engine
Tim Harmon, Rocketdyne (retired)

■ Noon – Break
Program resumes at 1 p.m.

AJ10-137 Apollo Service Module Engine
Clay Boyce,
Aerojet (retired)

Lunar Descent Engine
Jerry Elverum, TRW (retired)

Lunar Ascent Engine
Tim Harmon, Rocketdyne (retired)

Closing remarks
Steve Fisher, Technical Fellow, Rocketdyne

Appendix B

Speakers' Group Photograph

Pictured (left to right): Steve Fisher, Clay Boyce, Bob Biggs, Gerald Pfeifer, Tim Harmon, Gerard Elverum, Paul Coffman, and Shamim Rahman.

An actual F-1 Engine is shown in the background; a display model in front of the NASA SSC onsite visitor center, April 2006.

Appendix C

Robert Biggs' Presentation Viewgraphs

Appendix C

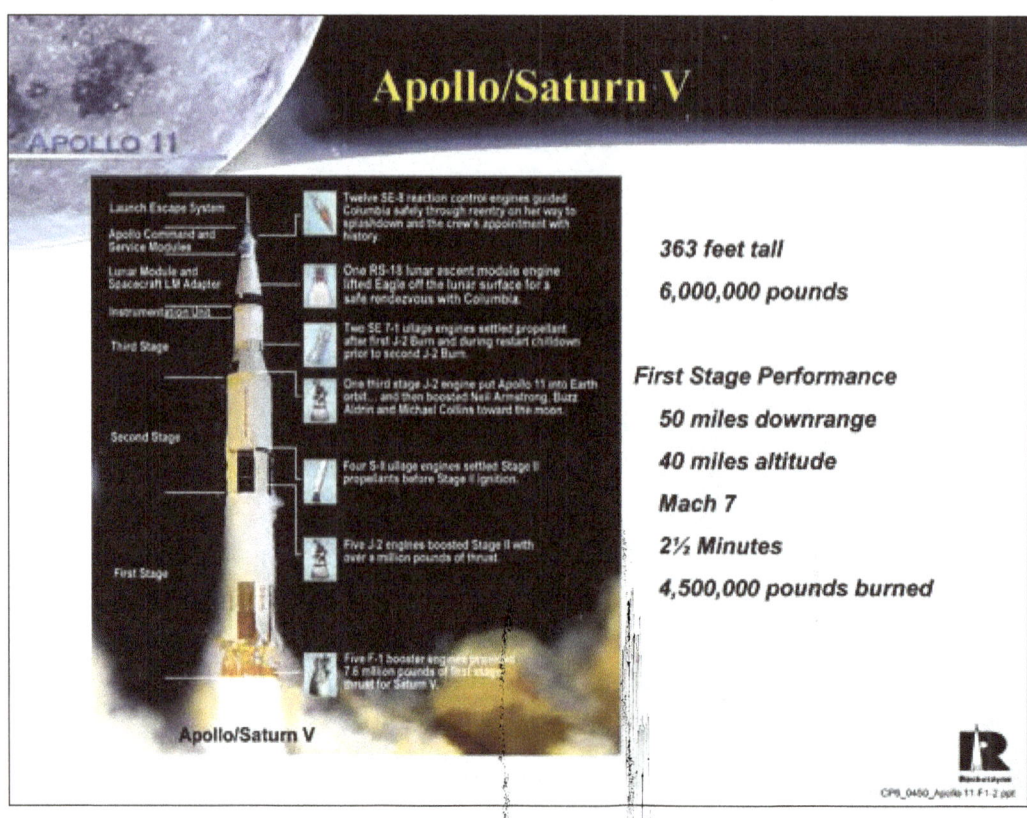

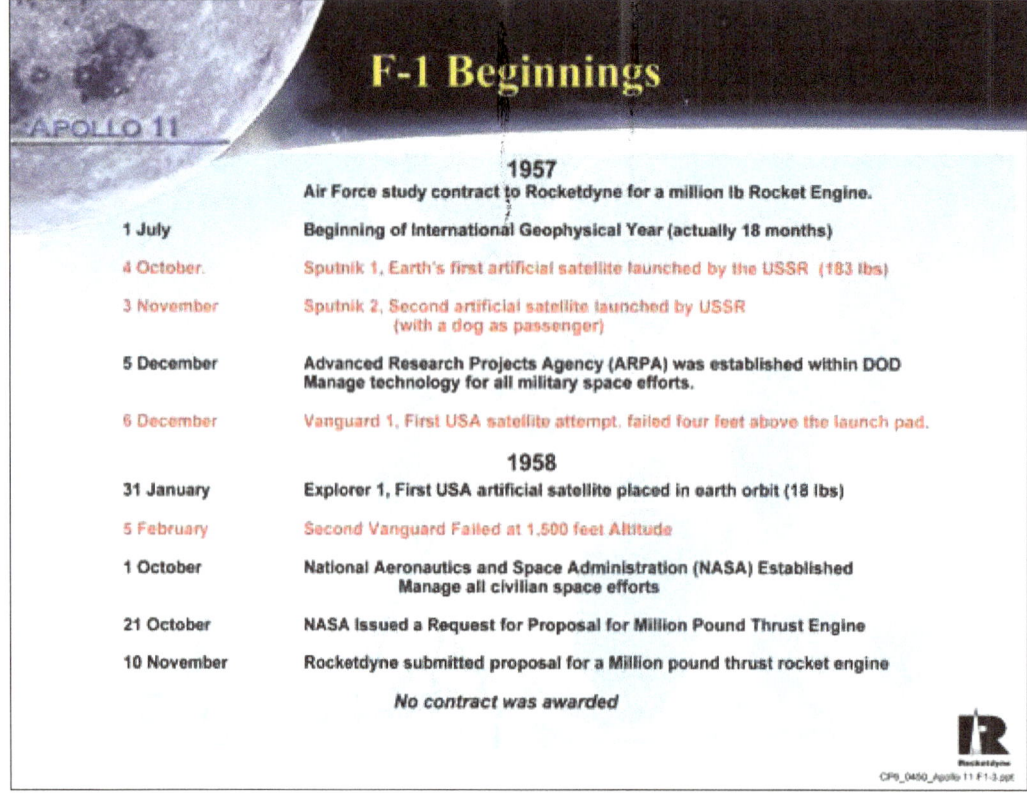

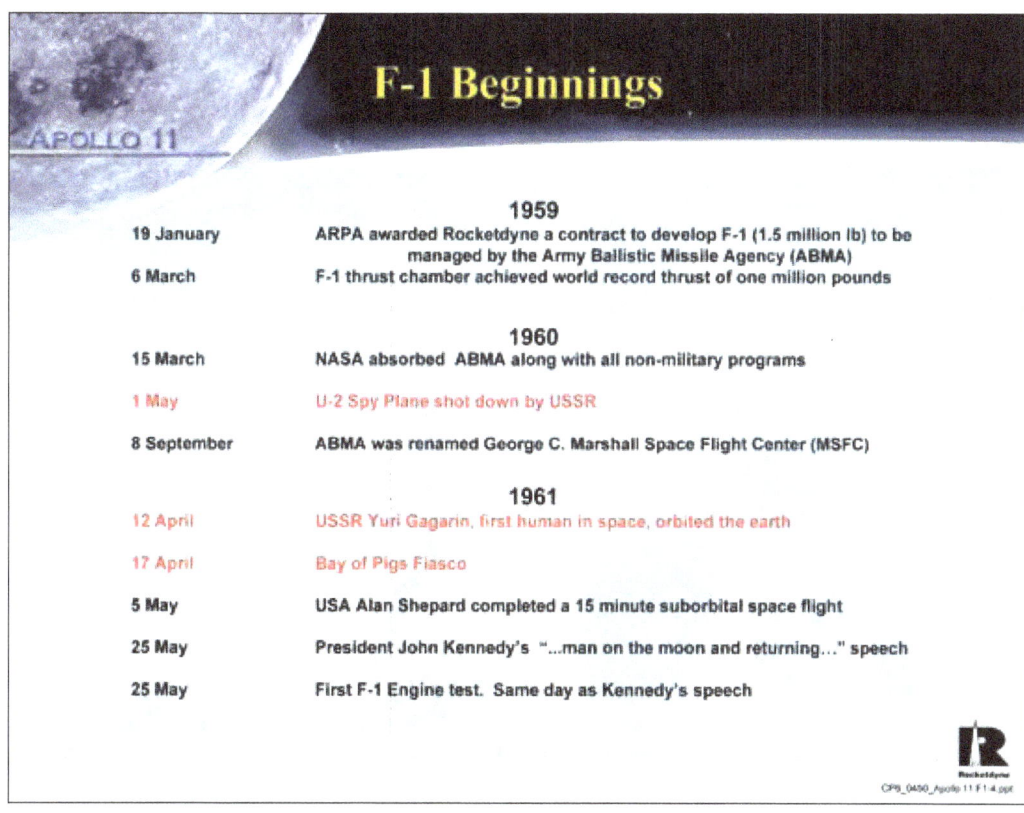

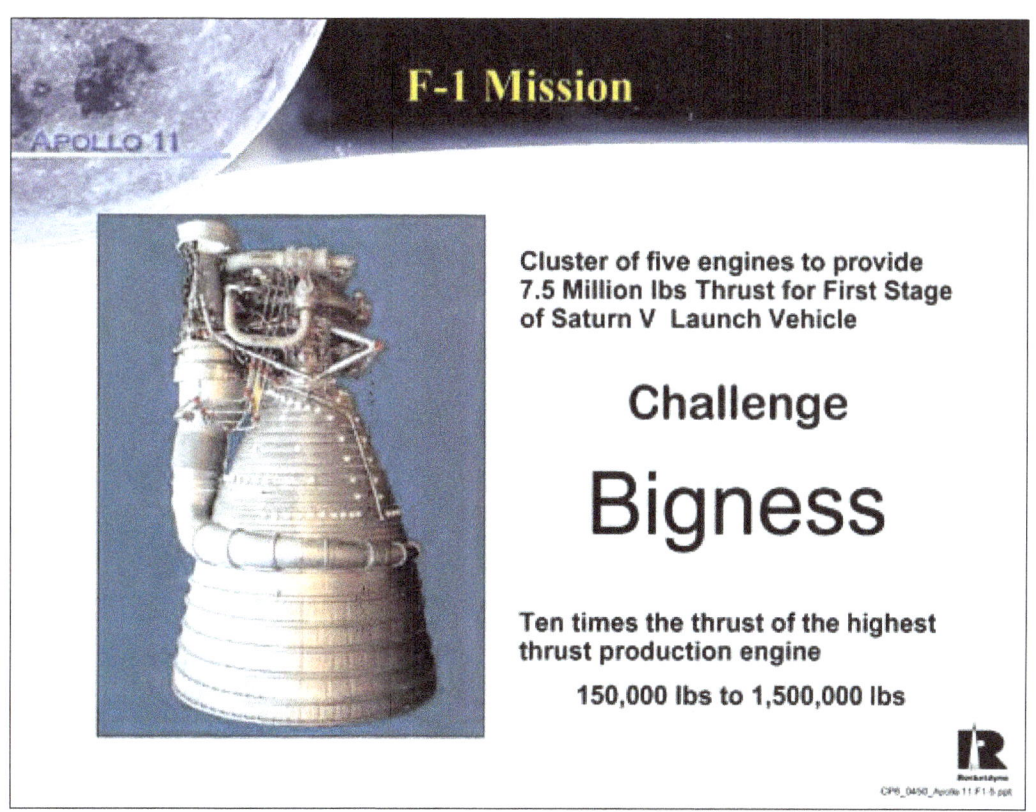

Appendix C

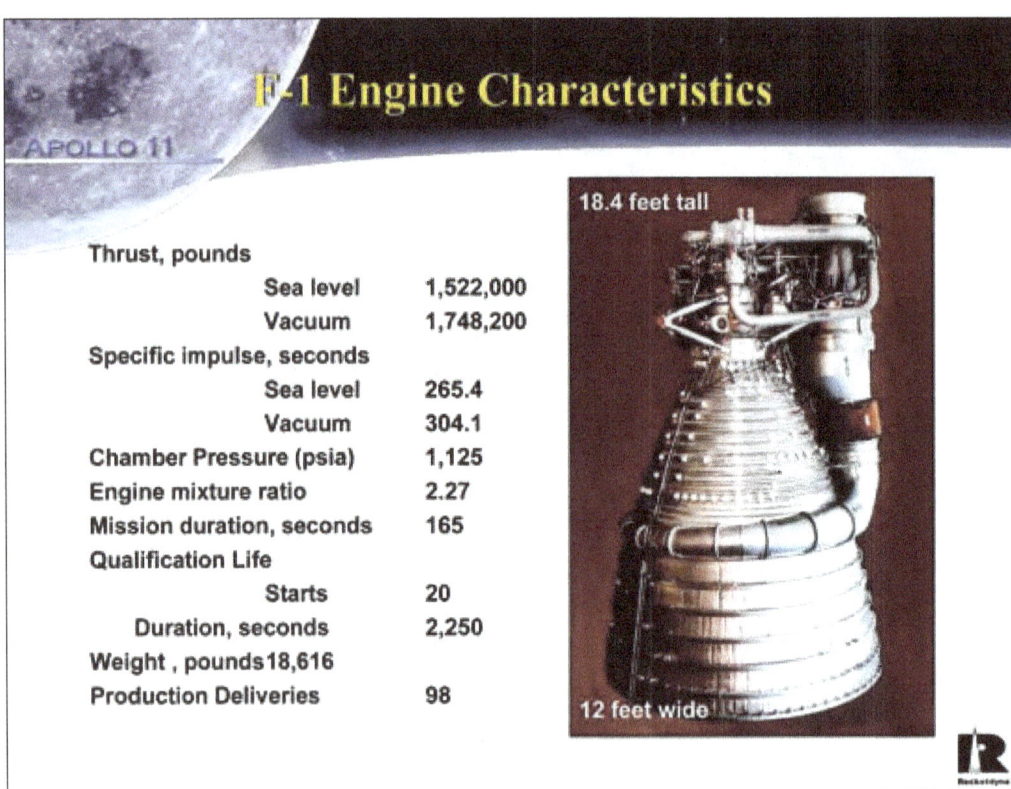

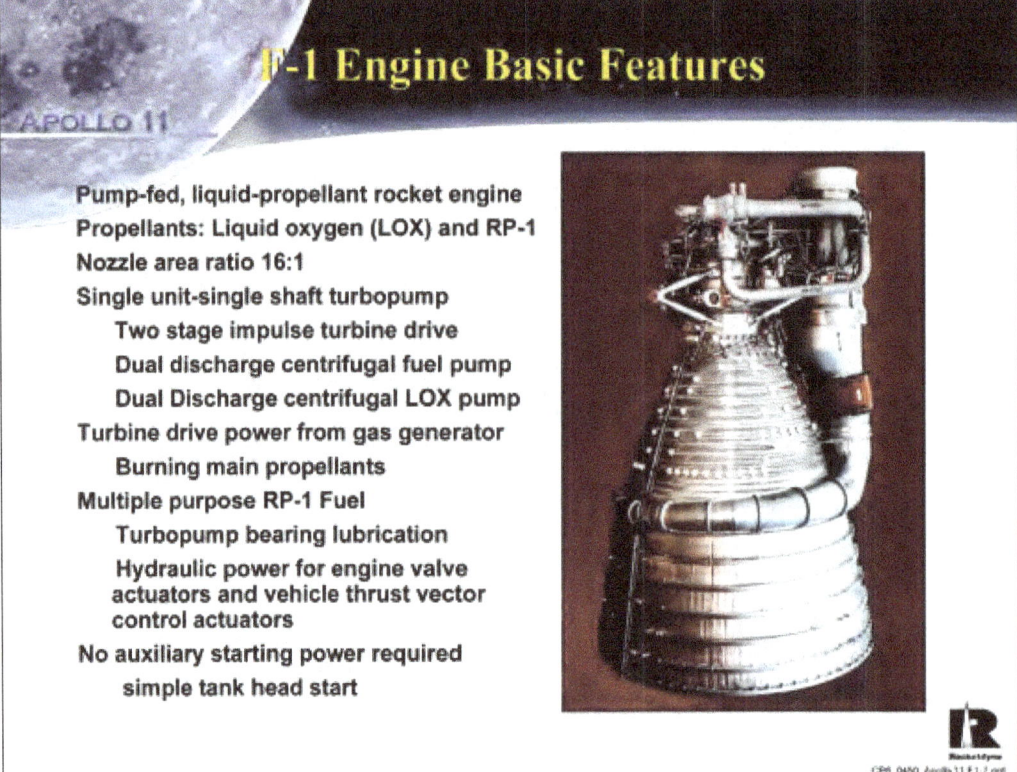

Robert Biggs' Presentation Viewgraphs

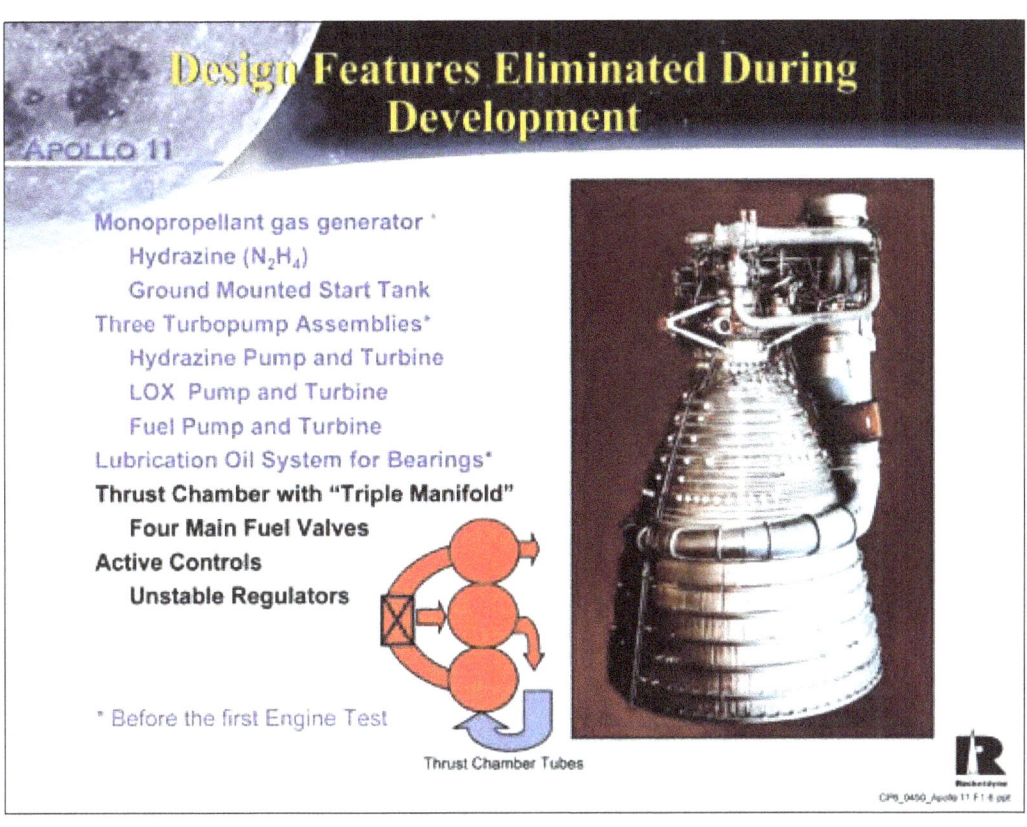

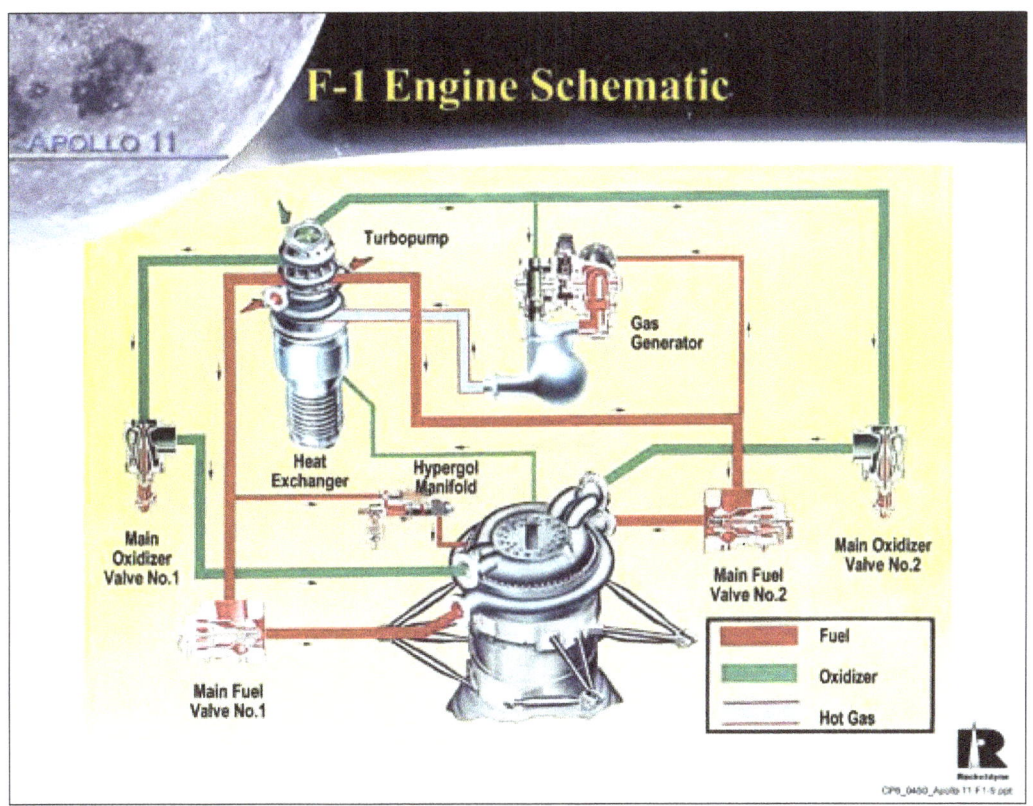

Apollo Rocket Propulsion Development 107

Appendix C

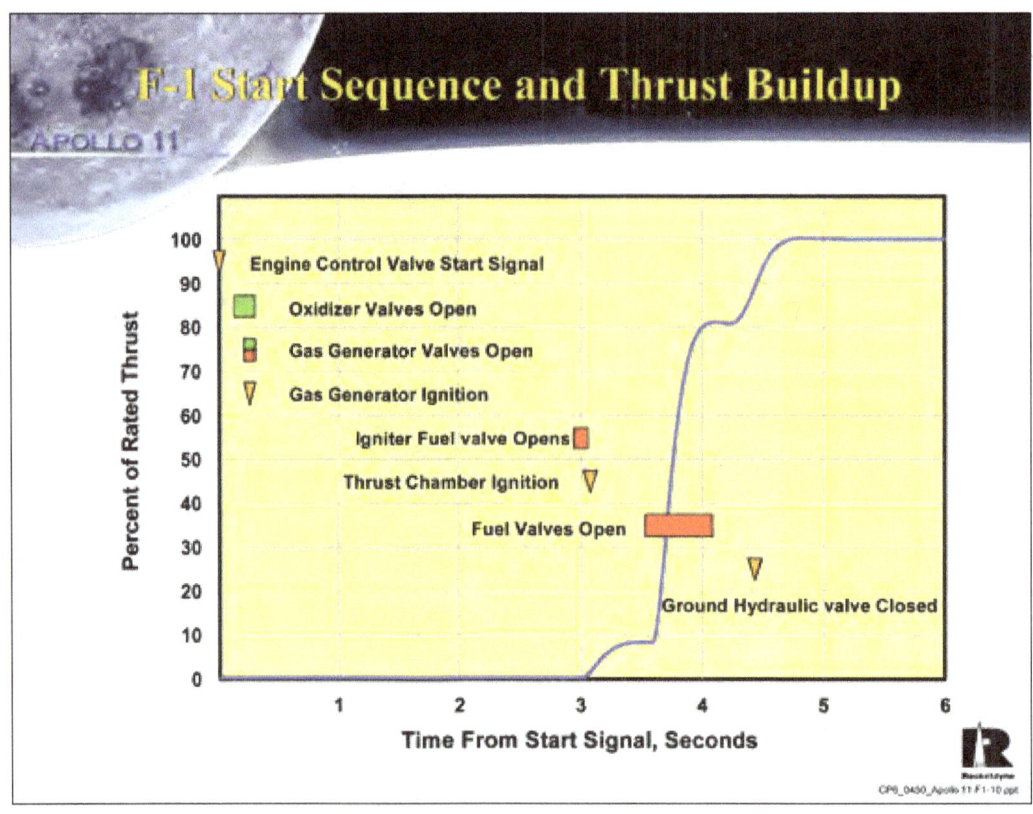

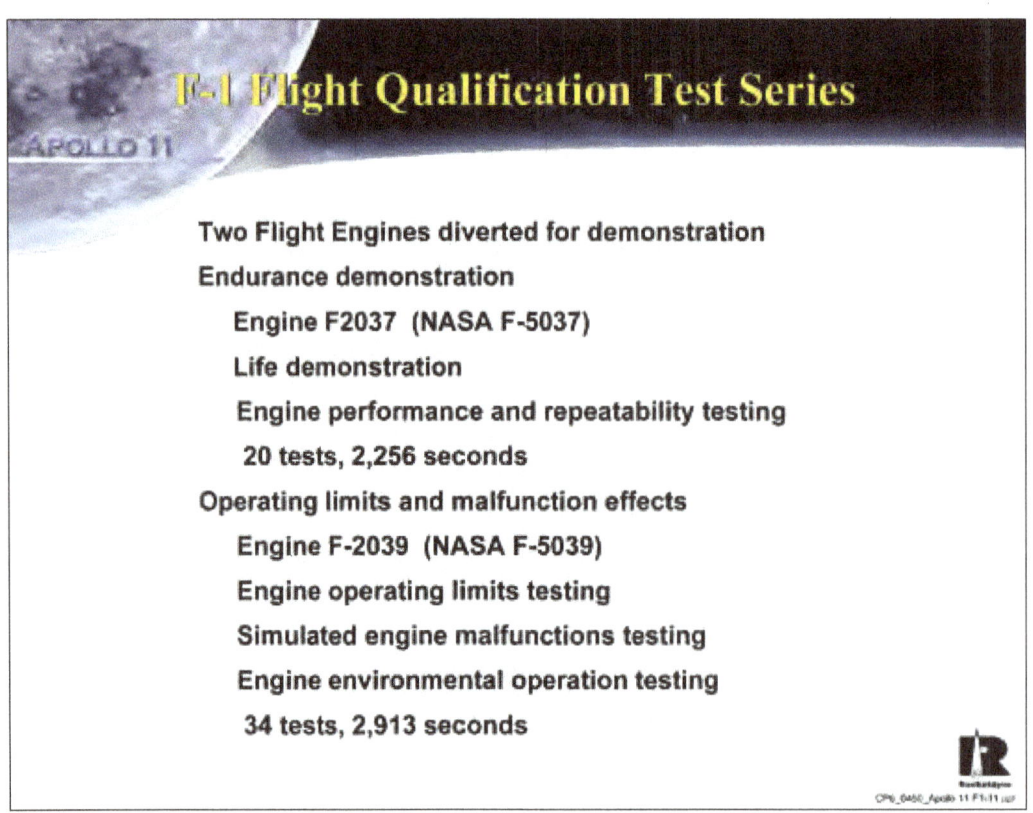

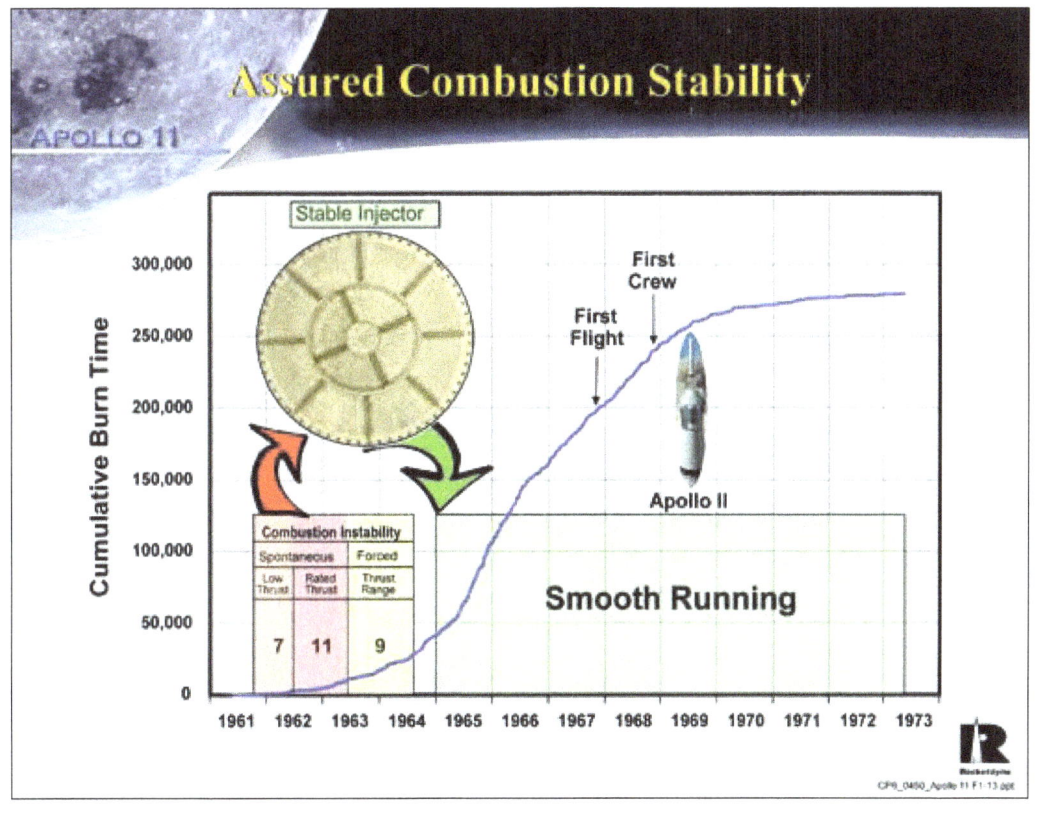

Appendix C

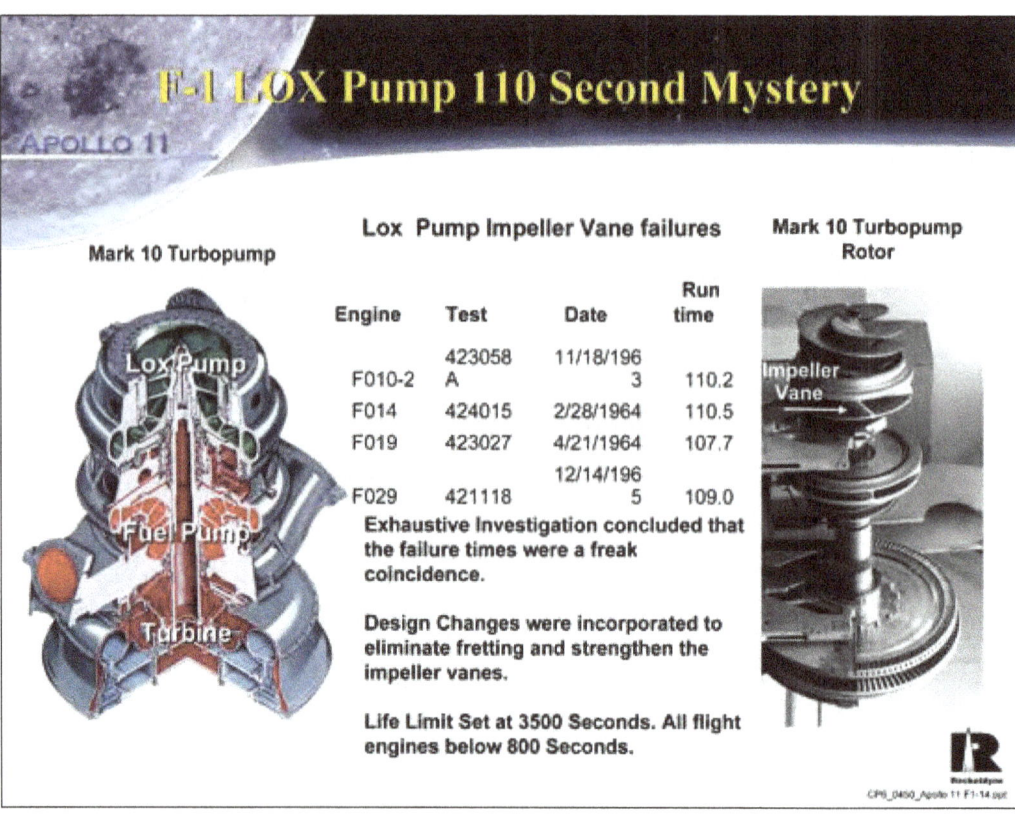

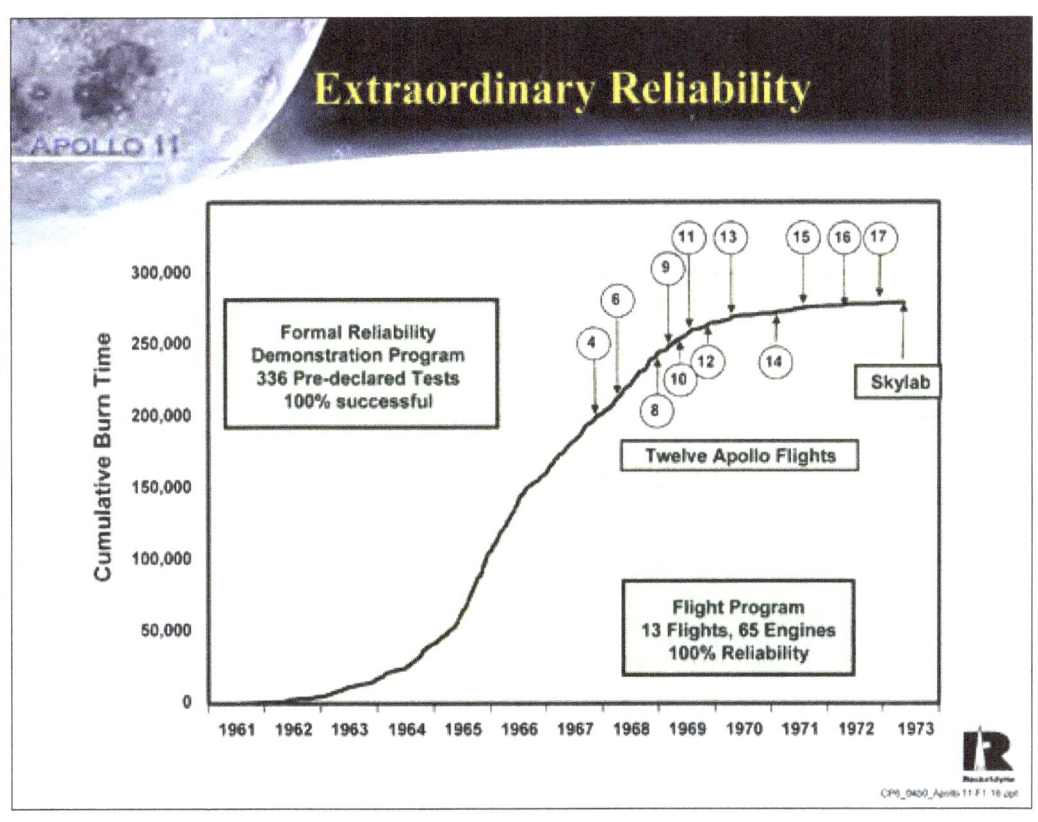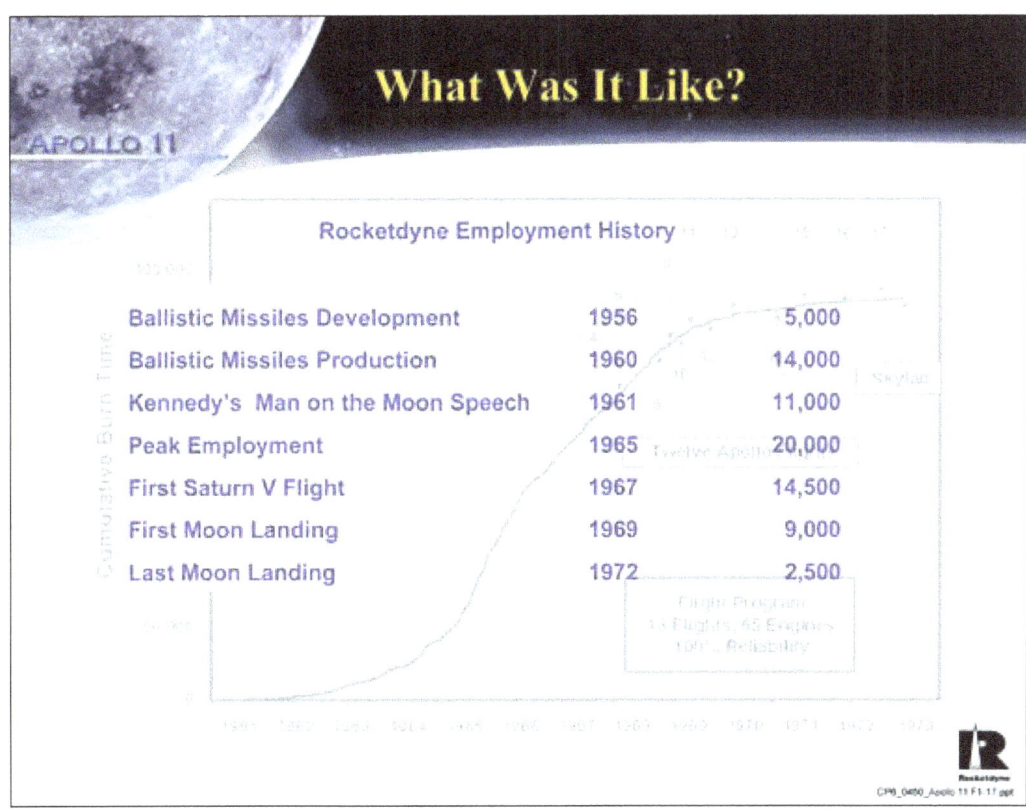

Appendix D

Paul Coffman's Presentation Viewgraphs

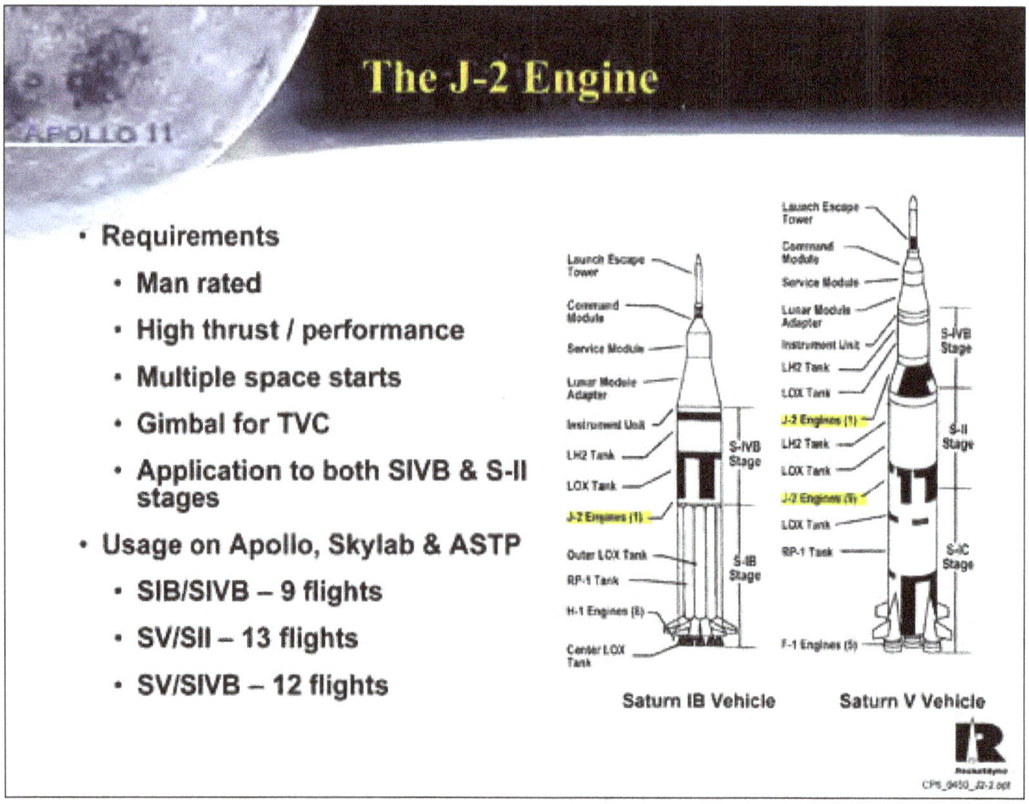

The J-2 Engine
APOLLO 11
• Requirements
• Man rated
• High thrust / performance
• Multiple space starts
• Gimbal for TVC
• Application to both SIVB & S-II stages
• Usage on Apollo, Skylab & ASTP
• SIB/SIVB – 9 flights
• SV/SII – 13 flights
• SV/SIVB – 12 flights
Launch Escape Tower
Command Module
Service Module
Lunar Module Adapter
Instrument Unit
LH2 Tank
LOX Tank
J-2 Engines (1)
Outer LOX Tank
RP-1 Tank
H-1 Engines (8)
Center LOX Tank
S-IVB Stage
S-IB Stage
J-2 Engines (5)
F-1 Engines (5)
S-II Stage
S-IC Stage
Saturn IB Vehicle
Saturn V Vehicle

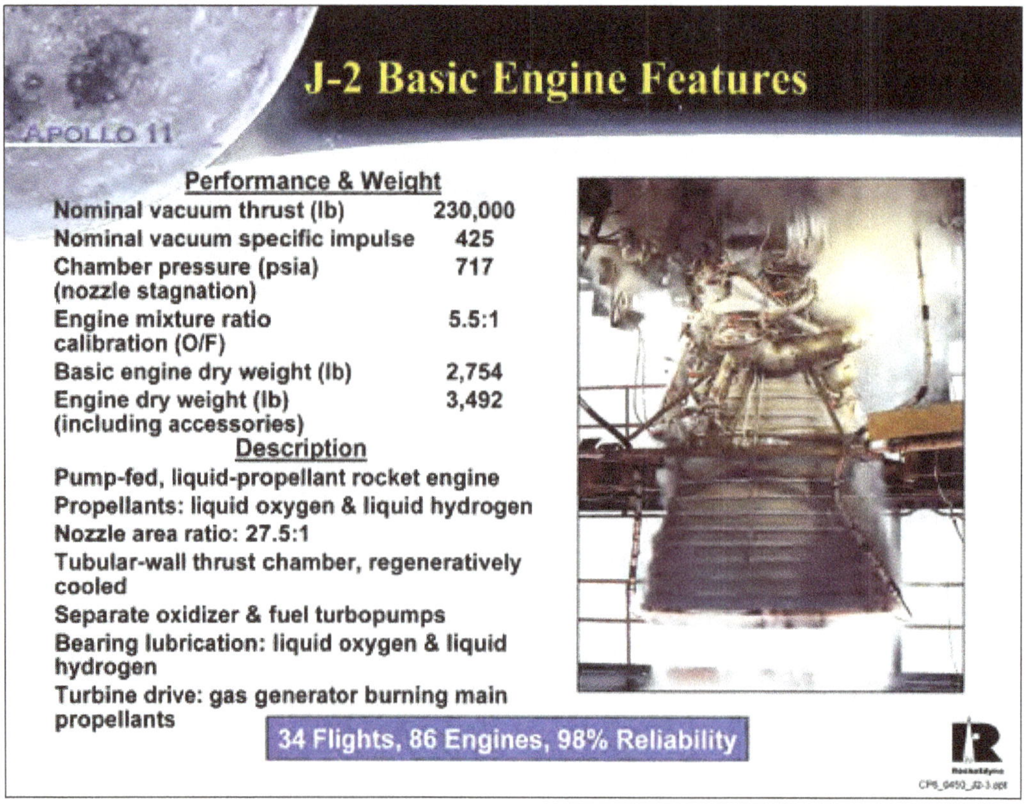

J-2 Basic Engine Features
APOLLO 11
Performance & Weight
Nominal vacuum thrust (lb) 230,000
Nominal vacuum specific impulse 425
Chamber pressure (psia) (nozzle stagnation) 717
Engine mixture ratio calibration (O/F) 5.5:1
Basic engine dry weight (lb) 2,754
Engine dry weight (lb) (including accessories) 3,492
Description
Pump-fed, liquid-propellant rocket engine
Propellants: liquid oxygen & liquid hydrogen
Nozzle area ratio: 27.5:1
Tubular-wall thrust chamber, regeneratively cooled
Separate oxidizer & fuel turbopumps
Bearing lubrication: liquid oxygen & liquid hydrogen
Turbine drive: gas generator burning main propellants
34 Flights, 86 Engines, 98% Reliability

Paul Coffman's Presentation Viewgraphs

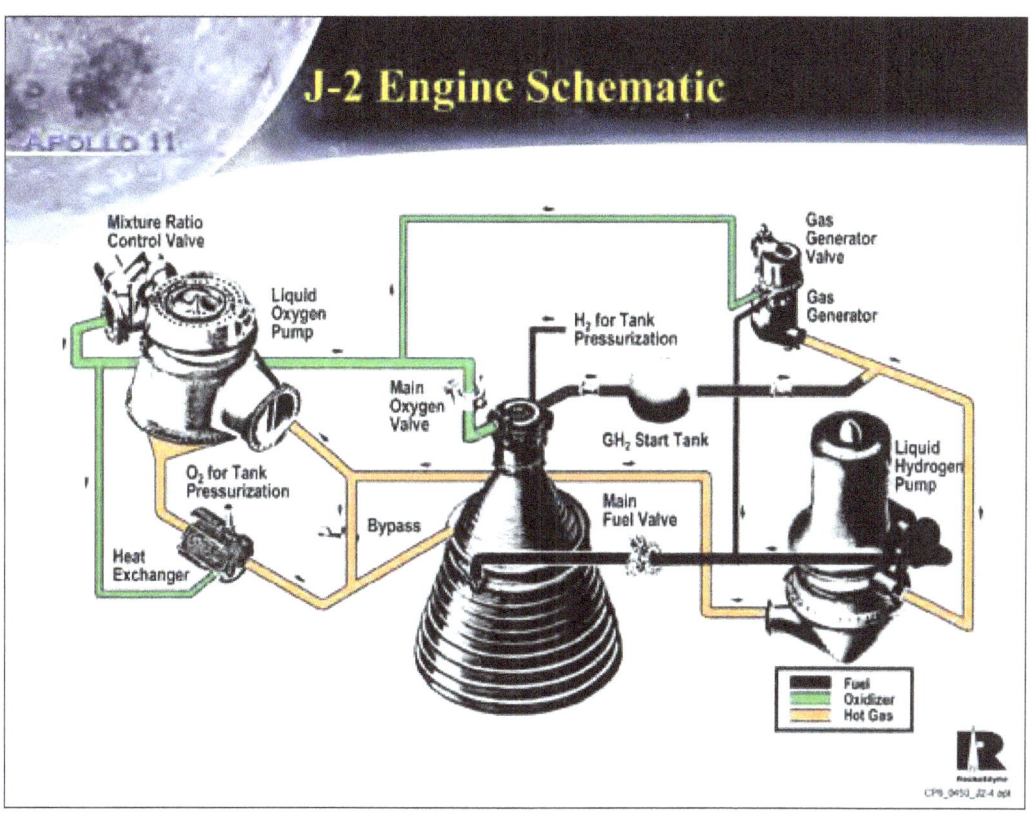

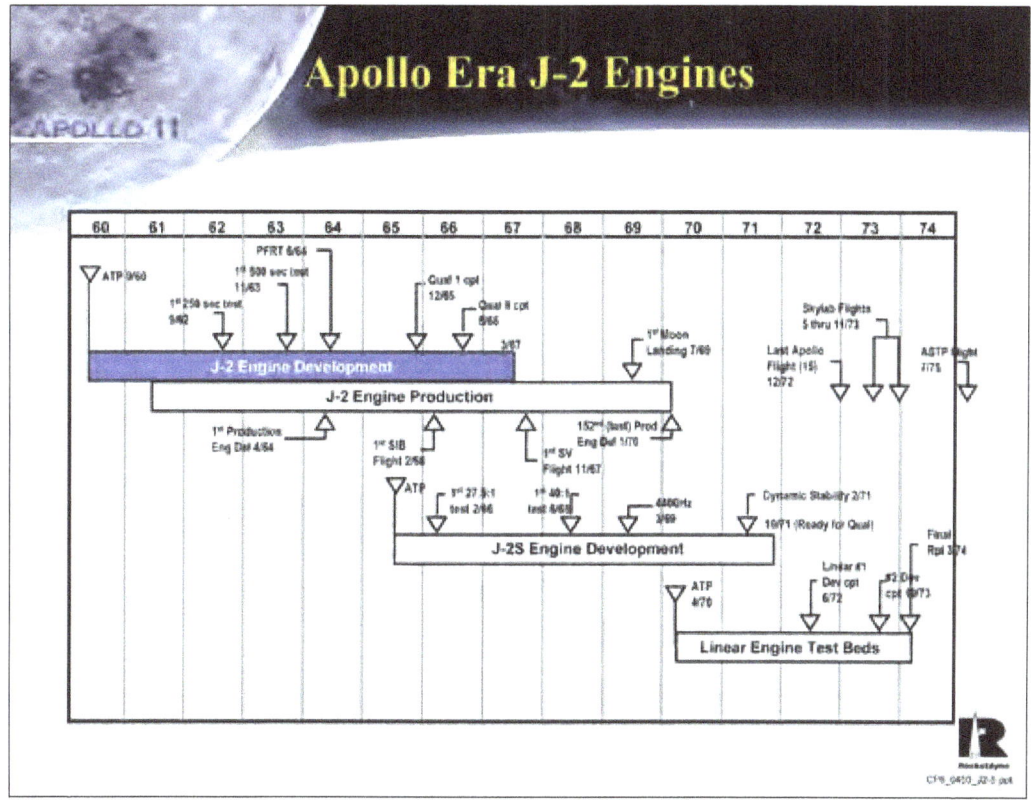

Apollo Rocket Propulsion Development 115

Appendix D

J-2 Engine Development

- Development engines - 38
- Engine tests – 1730 through qualification
- Engine (and stage) test facilities
  - Rocketdyne/ Santa Susana Field Lab
    - Including one (1) simulated altitude facility
  - NASA / MSFC
  - NASA / MTF (now SSC)
  - USAF / AEDC simulated altitude facility

J-2 Engine J2072 Qualification Program

	Requirement	Demonstrated
Endurance Tests		
Total Tests	30	30
Total Seconds	3750	3807.4
Total Duration At High Mixture Ratio	1500	1503.9
Operating Limits and Performance Tests		
Programmed Mixture Ratio; 470 Second Duration Tests	3	3
Gimbaling Pattern Cycles	8	8
Mixture Ratio Control Valve Calibration Point Tests	5	5
Mainstage Performance Tests	4	7
Heat Exchanger Pressurization Performance Point Tests	2	2
Hydrogen Tapoff Pressurization Performance Point Tests	1	1
Safety Limits Test	16	16
Start-restart Test Couples	2	2
Demonstration Of Restart Capability Tests	1	1
Start-stop Tests	As necessary	5
Altitude Simulation Test (Engine J-2073)	1	1

Selected J-2 Engine Development Issues

- Thrust chamber assembly ignition detection
- Engine start transient
- Engine sideloads
- AS 502 failure – ASI fuel flex line

J-2 Engine TCA Ignition Detection

- TCA ignition accomplished by ASI
  - Spark plugs
  - Hydrogen & oxygen
  - Ignition detectors developed
- Fusible link detector
  - Used on most ground tests, except at AEDC
  - Satisfactory reliability, but limited to single use
- Reusable IR detector
  - Used at AEDC for multiple testing
  - Unsatisfactory reliability for flight
- Result: ALL ground tests used ignition detector, NO flights used one!!

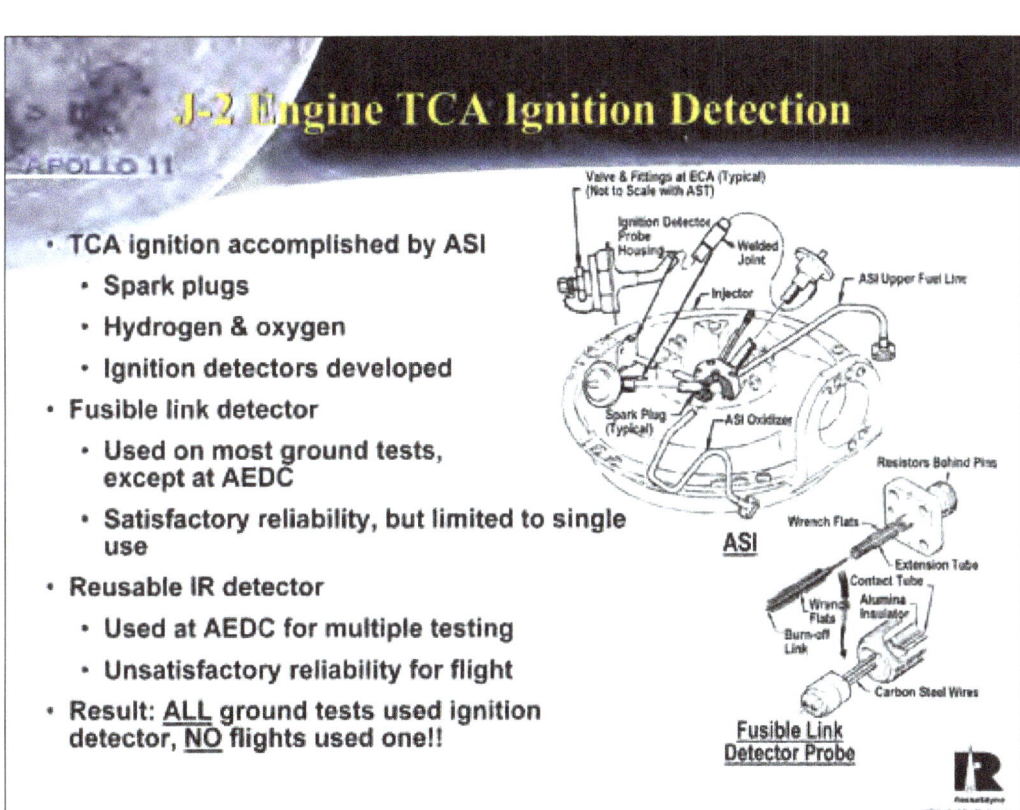

Appendix D

J-2 Engine Start Transient

- Engine start sensitivity – <u>THE</u> major J-2 development issue
  - Potential gas generator overtemp and fuel turbine burnout
- Contributing component characteristics:
  - Axial fuel turbopump stall
    - TCA thermal conditioning
    - Fuel lead time variation dependent on use
  - Complex main oxidizer valve pneumatic actuator operation
    - Two position
    - Extended ramp to full open
  - Turbopump start energy from high pressure cold GH_2
    - Controlled by start tank vent and relief valve
    - Secondary Helium temperature effect
  - Extensive thermal conditioning and sequence control required

J-2 Start Transient Development

- Initial problems encountered during ground test
  - SSFL and "sea level" test sites
  - Simulated altitude testing at SSFL/VTS-3A addressed the issue
    - Small capsule
    - Steam driven hyperflow ejector
  - Early S-IB flights indicated problem <u>NOT</u> resolved (or well understood)
    - Additional simulated altitude test conducted at AEDC/J-4
      - Large vacuum facility
      - Extensive analytical modeling capability developed
        - Facilitates modern engine development

J-2 Start Transient Control Achieved

- Satisfactory starts on all ground and flight applications
  - Relatively slow (3.5 – 4 sec to 90% P_c)
  - Repeatible, with narrow bandwidth
- Analytical model development

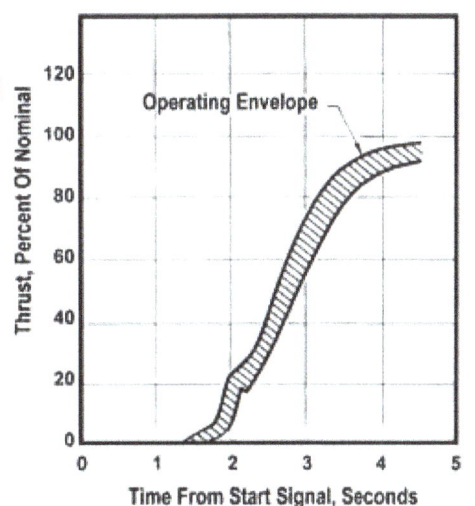

J-2 Engine Sideloads

- Unique J-2 Thrust chamber nozzle configuration
  - Adverse pressure gradient (APG)
    - Enable "sea level" operation with full flow 27.5:1 nozzle
    - Facilitate development
  - Acceptable minor performance loss
    - Analysis
    - Model hot fire test
- Unstable separation and high sideloads experienced
  - During start transient operation
  - During low thrust (PU) operation
  - Thrust chamber damage incurred
    - Actuator arm attach points (>100Klb loads)
    - Nozzle deformation

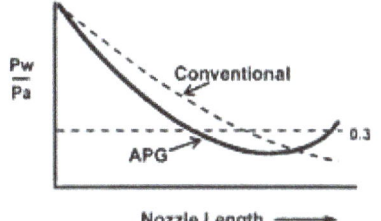

J-2 Engine Sideload Resolution

- Water cooled bolt-on "diffuser" developed
  - Classic diffuser >20ft long!
  - Subscale models (wood & metal) tested
  - Uncooled full scale verification
    - Six (6) inch length
- Strengthened nozzle
  - Tubing welded into hat bands
  - "T-ring" bolted to exit
- Restrained thrust chamber
  - Side Load Attach Mechanism (SLAM)
  - Release to allow gimbal operation

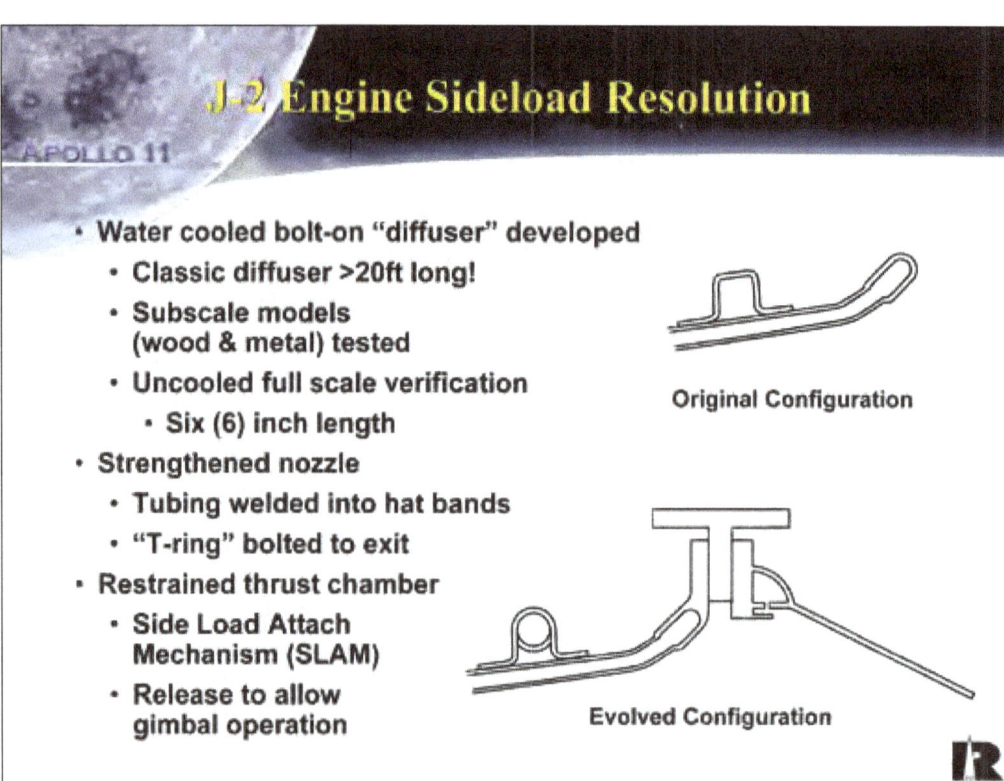

Original Configuration

Evolved Configuration

J-2 Engines on AS 502

- AS 502 launched April 4, 1968
  - Second launch of Saturn V vehicle (unmanned)
  - One of the five SII engines cutoff prematurely
  - The SIVB engine did not restart after its initial burn & nominal coast
- Fight data analysis showed significant anomalous operation
  - Similar failure mode projected for <u>both</u> engines!
- Expedited effort to:
  - Verify observed anomalies with ground test simulations
  - Define the failure mechanism analytically
  - Incorporate corrective action for subsequent Saturn flights

Engine J2042 on AS 502 SIVB Stage

- Engine operation nominal during pre-launch operation (704.6 sec mainstage)
  - Five engine and two stage acceptance tests
- Flight data anomalies observed
  - Engine area temperatures
    - Chilling at 645 sec
    - Heating at 696 sec
    - Chilling at restart
  - Performance decay
    - Starting at 684 sec
  - Restart failure despite
    - Proper conditions
    - Valve operation
- Failure mode hypothesis
  - Small ASI fuel bellows leak at 645 sec, with failure at 696 sec
  - ASI burnout

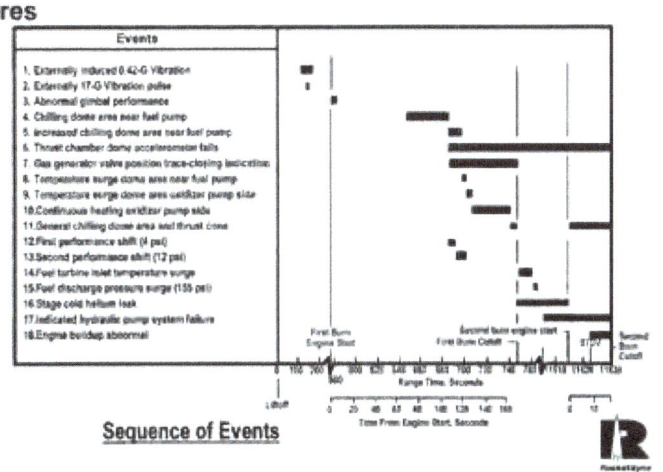

Sequence of Events

Failure Simulation

- Engine J016-4 configured to simulate J2042
  - Test equipment added to allow ASI flow control & overboard dump
  - Sequence of events planned:
    - 65 sec of normal operation
    - 35 sec of 0.6 lb/sec ASI fuel leak
    - Increase leak to complete line failure
    - Backflow of ASI combustion products for 29 sec
- Posttest evaluation
  - Substantial main injector ASI cavity erosion
  - ASI burnout
  - Performance loss similar to J2042 experience
- Similar test demonstration duplicated SII anomaly

Appendix D

Analysis and Resolution

- AS 502 engine ASI fuel line Assemblies had single ply metal bellows
  - Analysis indicated:
    - Flow induced vibration could cause bellows failure
    - Liquid air "damping" would prevent failure on ground test
  - Flow tests verified analysis
    - Inert fluids
    - Liquid hydrogen
  - Design modifications
    - Replacing bellows section with solid line
    - Incorporation of Class I welds on ASI fuel line assembly
  - Revised configuration qualified and incorporated on all engines
    - Returned to flight (manned !) in just over 8 months (AS 503)
    - No problem recurrence

J-2 Engine Development Summary

- Began in 1960 as part of the national goal
  - Schedule driven
  - Low technology readiness in many areas
- Accelerated development accomplished
  - First long duration engine test (250 sec) in two years
  - Initial formal test demonstration in four years
  - Formal development (QUAL II) completed in six years
    - "Flight support" testing continued
    - J-2S & Linear engine development through 1973
  - Over eighty different unplanned engine cutoff modes experienced
    - Prompt resolution enabled Apollo mission successes
  - J-2 engine experience enabled subsequent programs
    - SSME
    - RS-68
    - XRS-2200 (X-33)

Appendix E

G. R. "Jerry" Pfeifer's Presentation Viewgraphs

Appendix E

Origin of Apollo RCS 100 Lbf Thruster
AEROJET

- Based on IR&D work performed by Warren Boardman (retired), Jerry Pfeifer (Boeing Rocketdyne) et.al starting in the 1960 time period.
  - Two thrust levels (5 and 25 Lbf)
  - single unlike impinging injector doublet
  - Molybdenum combustion chamber (Durak (MoSi$_2$ Coating)
  - Eckel fast response solenoid valves

Advent 25 lbf engine was life tested using Molybdenum chamber in 1962.
- Single unlike doublet Injector
- Bolted assembly

R-1E-1 (Without Valves)

5 Lbf Engine

5 LB. THRUST SYNCOM II ENGINE

April 25, 2006

Original Apollo Configurations
AEROJET

- NASA/Rockwell funded Marquardt for a "Common" engine to be used on the Service Module, Lunar Module and Command Module (buried)

Early Model R-4 100 lbf Engine
One Piece Molybdenum Chamber

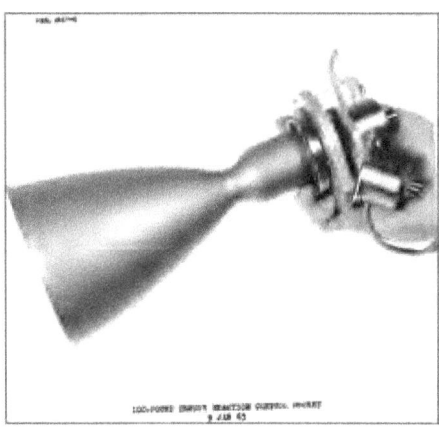

A R-4D Concept that never Flew
A Good Idea that Lost to Politics
The Cold Wall - Not the Berlin Wall

April 25, 2006

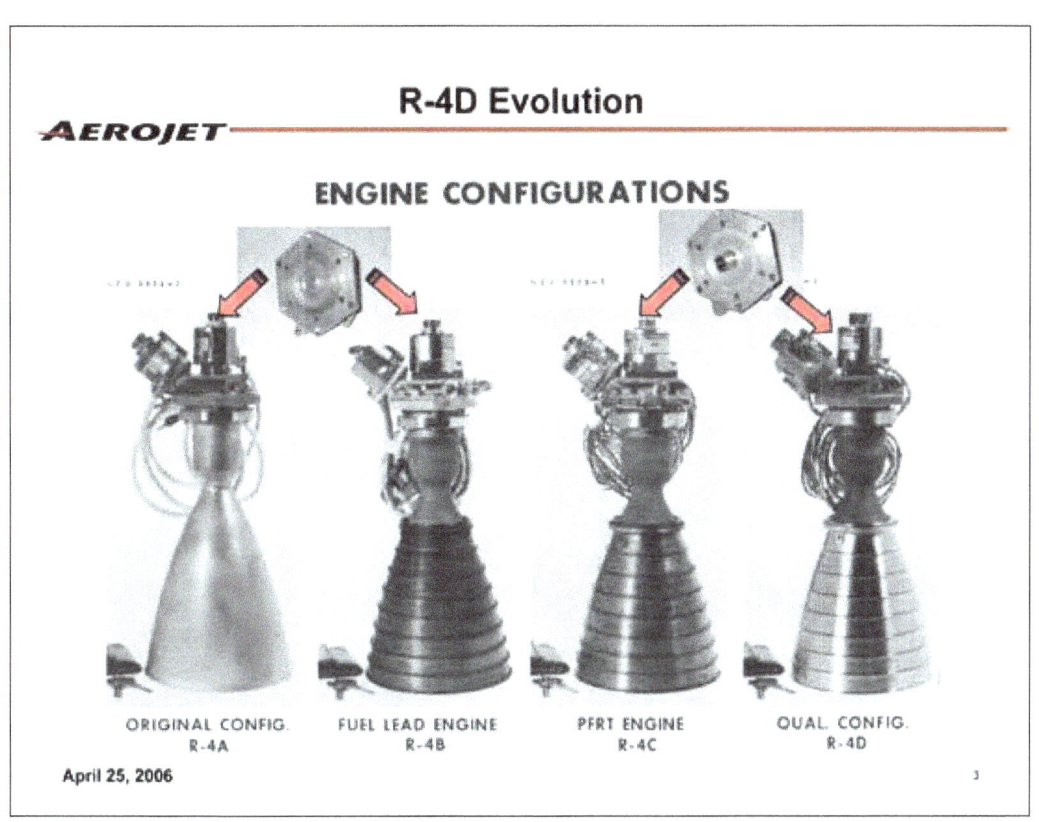

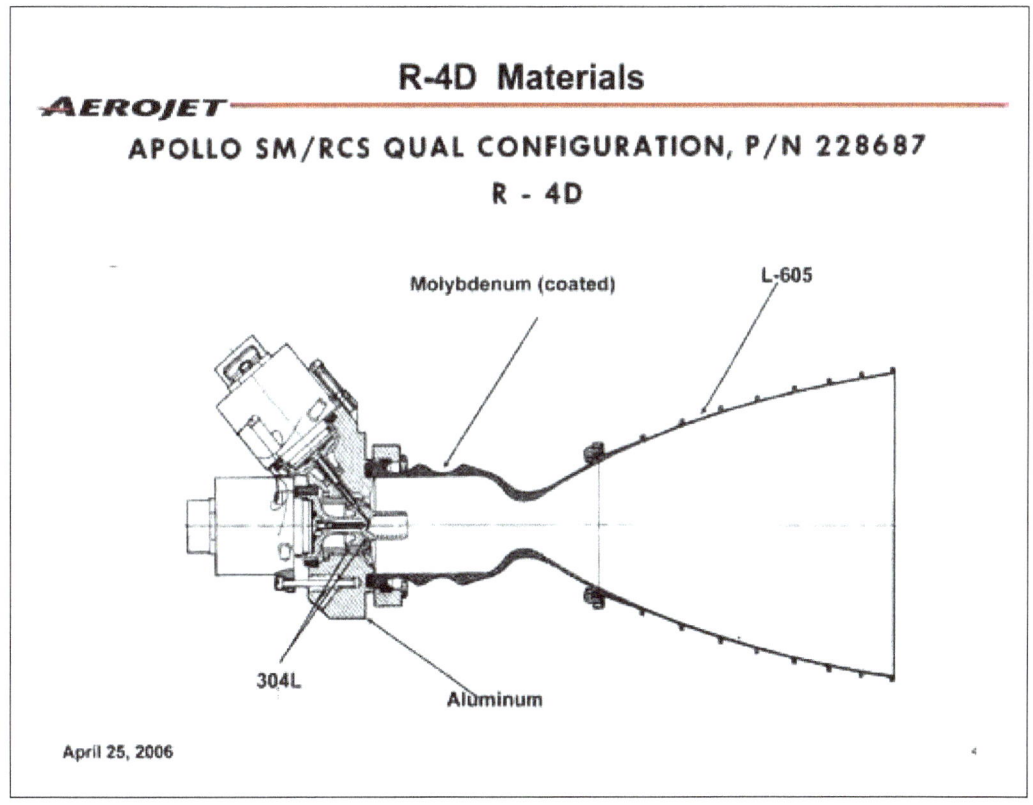

First Use of R-4D Apollo engine - Lunar Orbiter

LUNAR ORBITER SPACECRAFT
ASSEMBLY OF UNIT NO. 2

April 25, 2006

Problems

- HHTB - High Heat Transfer Burning - a.k.a. 1st tangential combustion instability
    - Went away never to return - after the pre-ignition chamber was integrated.

- Spikes - The result of detonating (by a vapor phase ignition delay) residual unburned and partially reacted propellants that deposited on the chamber wall
    - At 0.010 Electrical pulse widths the chamber cooled down due to the propellant holdup in the manifold
    - Eliminated by heating the injector to 70°F (with MMH) and 125°F (with A-50)
    - SP-4205 statement that "Marquardt eliminated spiking by installing a small tubular "pre combustion" chamber inside the engine" is in error.

- ZOTS (inner-manifold explosions)
    - Resulting from operation in a pressure field above the triple point of MMH or A-50 and is aggravated by gravitational field.
        - It happens at sea level up to 70,000 ft or so.
        - Only occurs in space if there is a significant fuel leak that pressurizes the chamber between firings (pulse mode).

April 25, 2006

- Ignition characteristics – "Spikes"

AEROJET

- The development of the reaction control engines for the Apollo LM and SM resulted in additional development and testing because of premature disassembly of engines during extended duty cycle pulse mode testing at low temperatures.
- The molybdenum combustion chamber would disintegrate during pulse mode firings where the equivalent propellant mass contained in the injector manifold exceeded the operating time of the pulse (about 10 milliseconds)
- The emptying process included condensation of the partially reacted and unburned neat propellants on the interior walls of the combustion chamber (nitrates).
- During the subsequent ignition (if the chamber was cold) the ignition delay could result in a "vapor phase detonation or "spike" that would act as an explosive source for the detonation of the nitrates
- The rest is history

April 25, 2006

Ignition characteristics – "Spikes"

AEROJET

- Elimination of this undesirable operational characteristics was accomplished by maintaining the injector heat above 70°F (120°F for Aerozine-50 fuel)
- Additional margin was subsequently accomplished by the incorporation of a niobium material combustion chamber which has low temperature ductility and can absorb the localized high strain rate detonations
- A detailed discussion of this phenomena is documented in:

   "Ignition Transients in Small Hypergolic Rockets" Juran and Stechman

   Journal of Spacecraft and Rockets, Volume 5, Number 3, Pages 288-292,

   March 1968

April 25, 2006

Appendix E

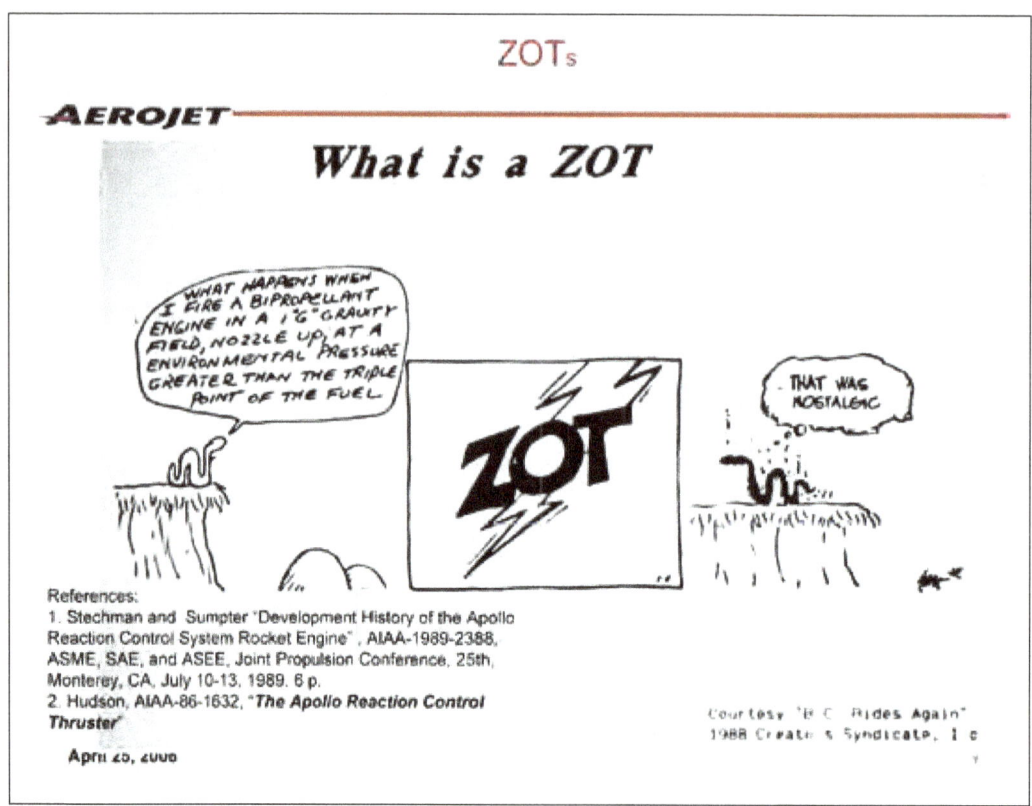

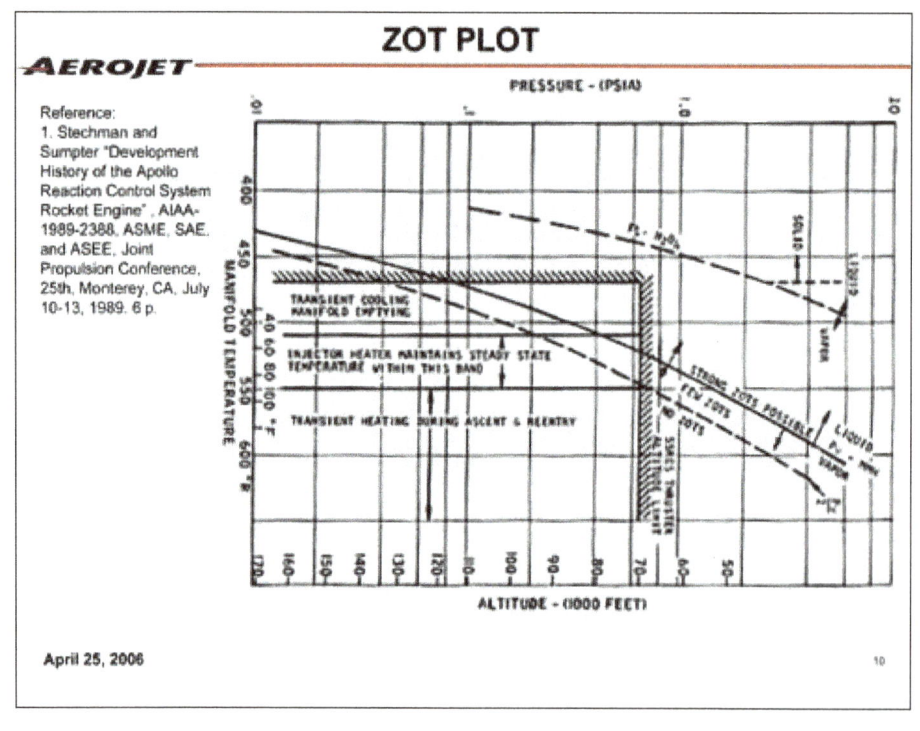

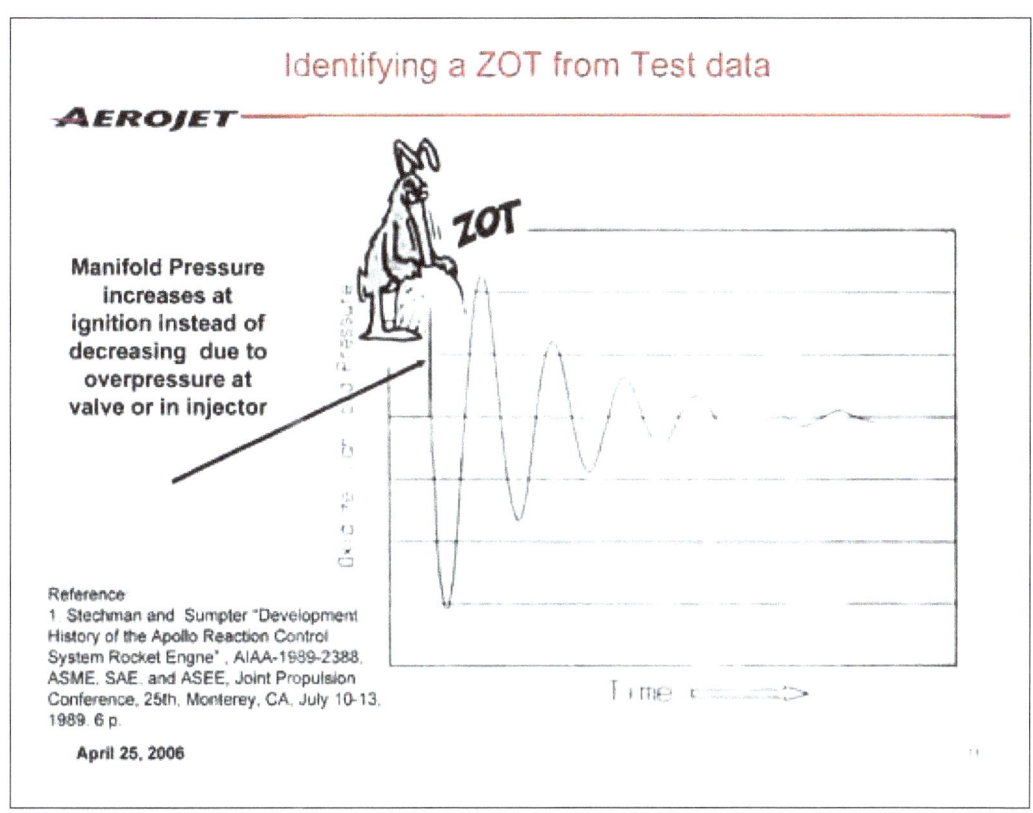
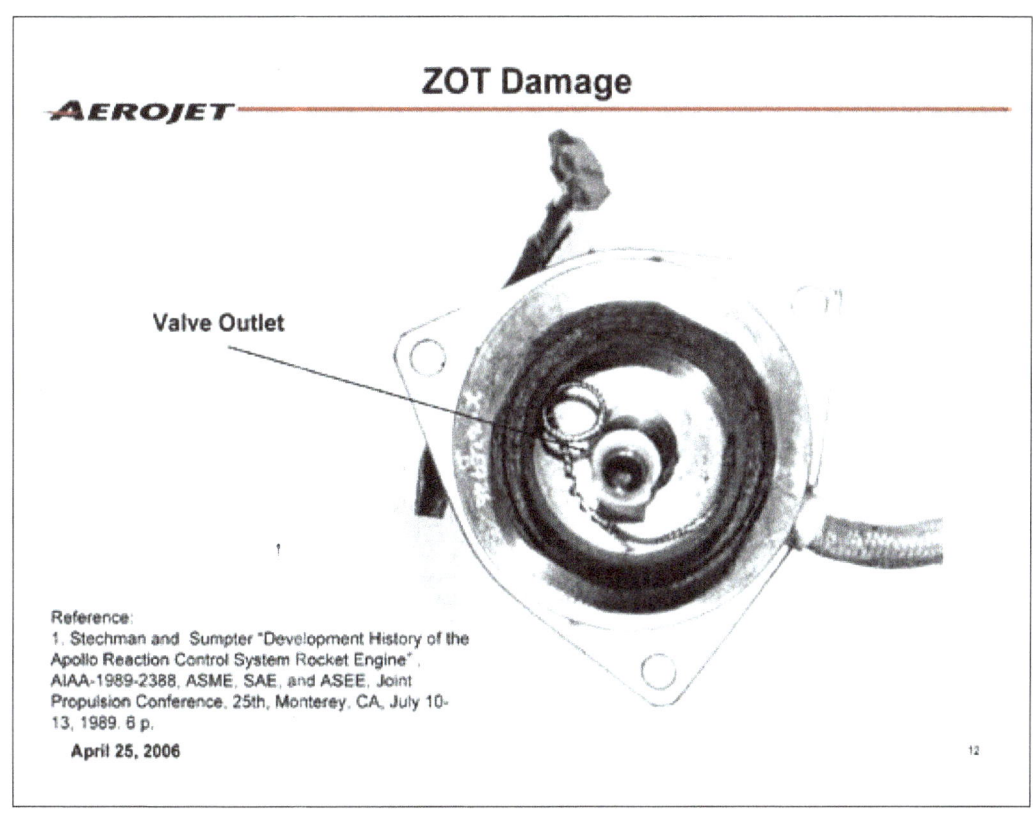

Appendix E

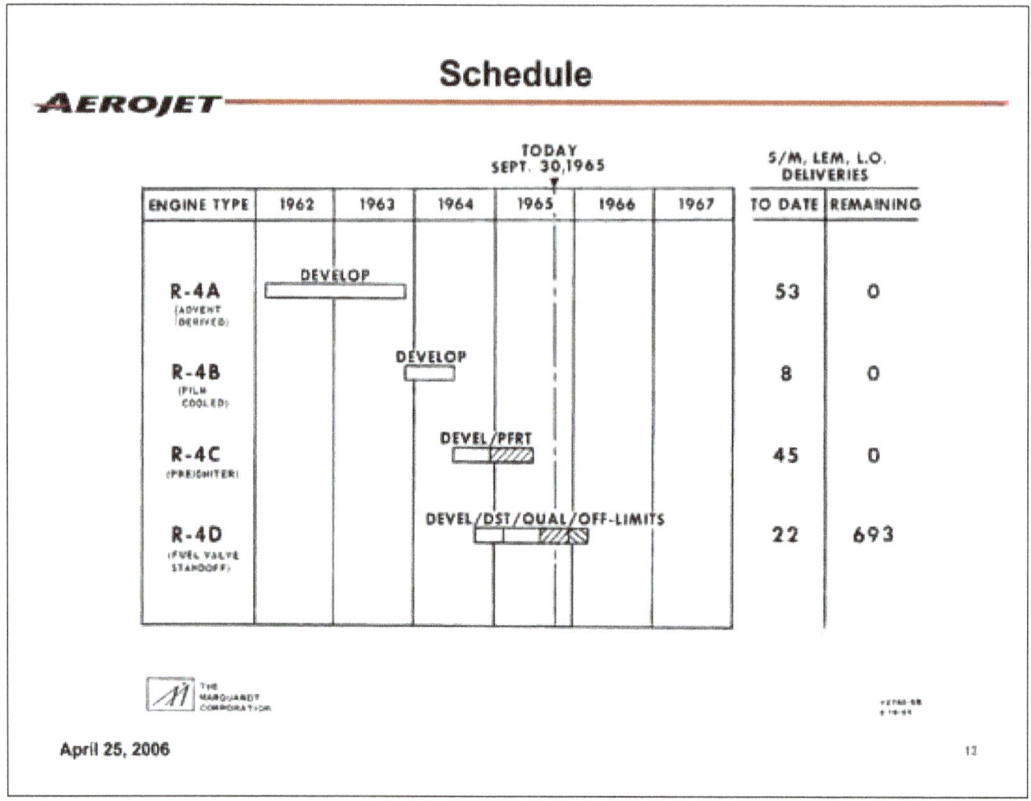

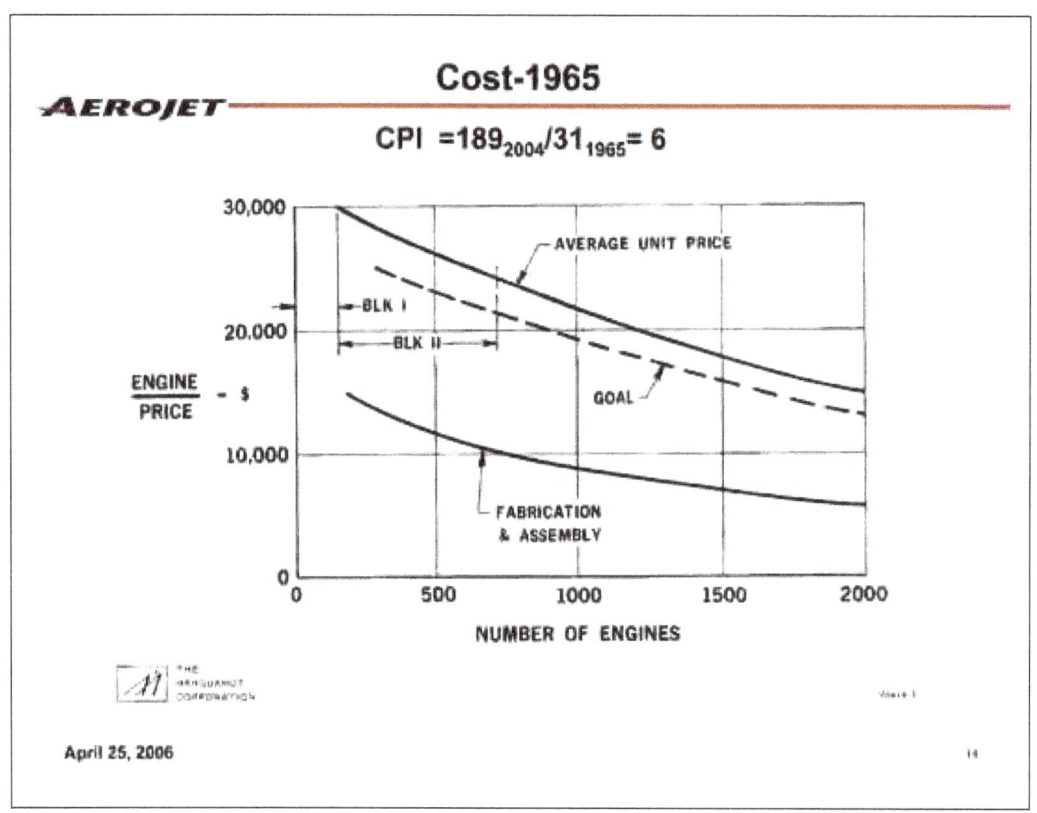

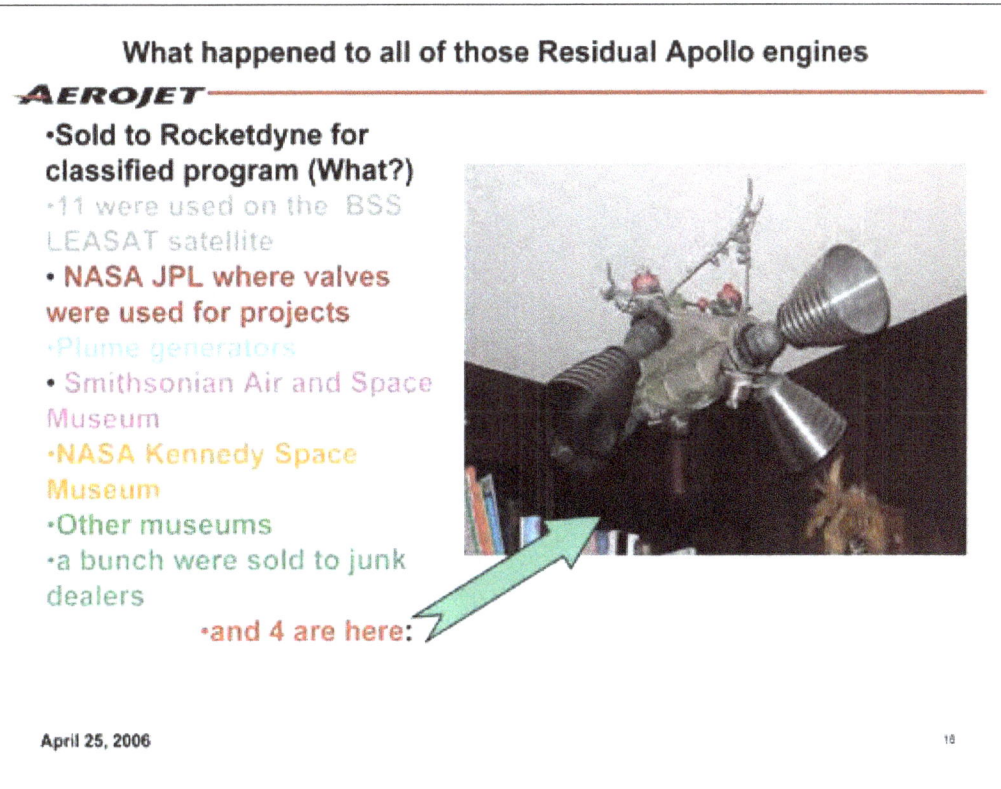

Appendix E

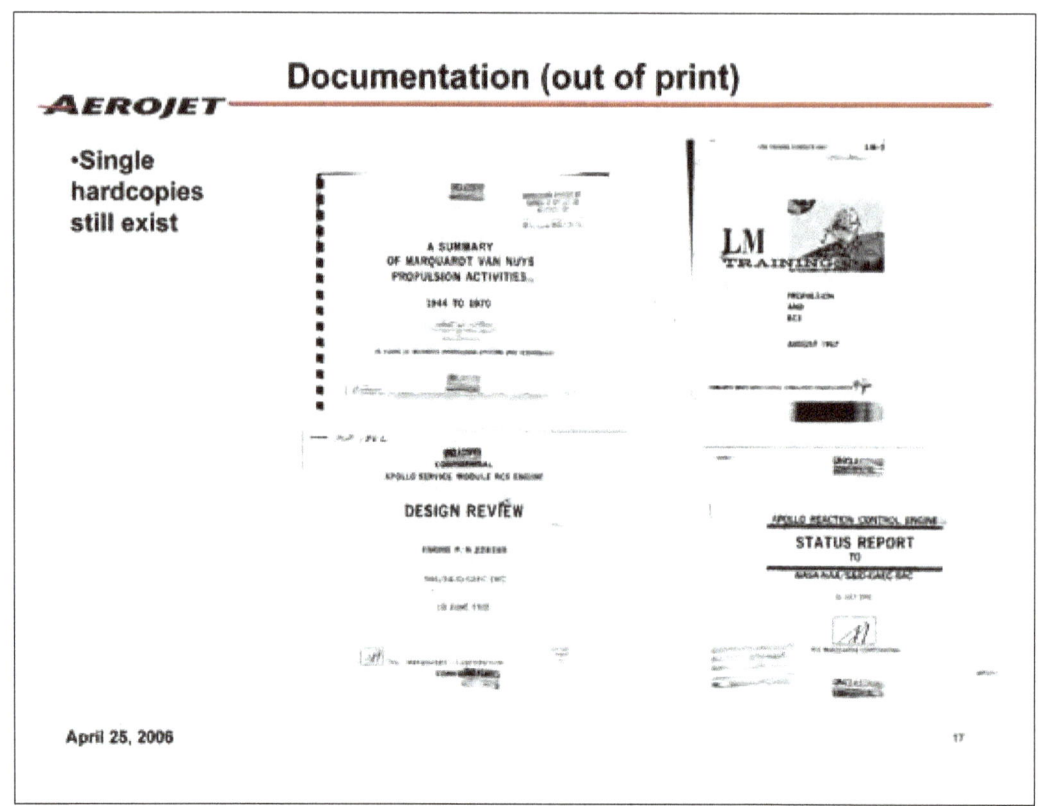

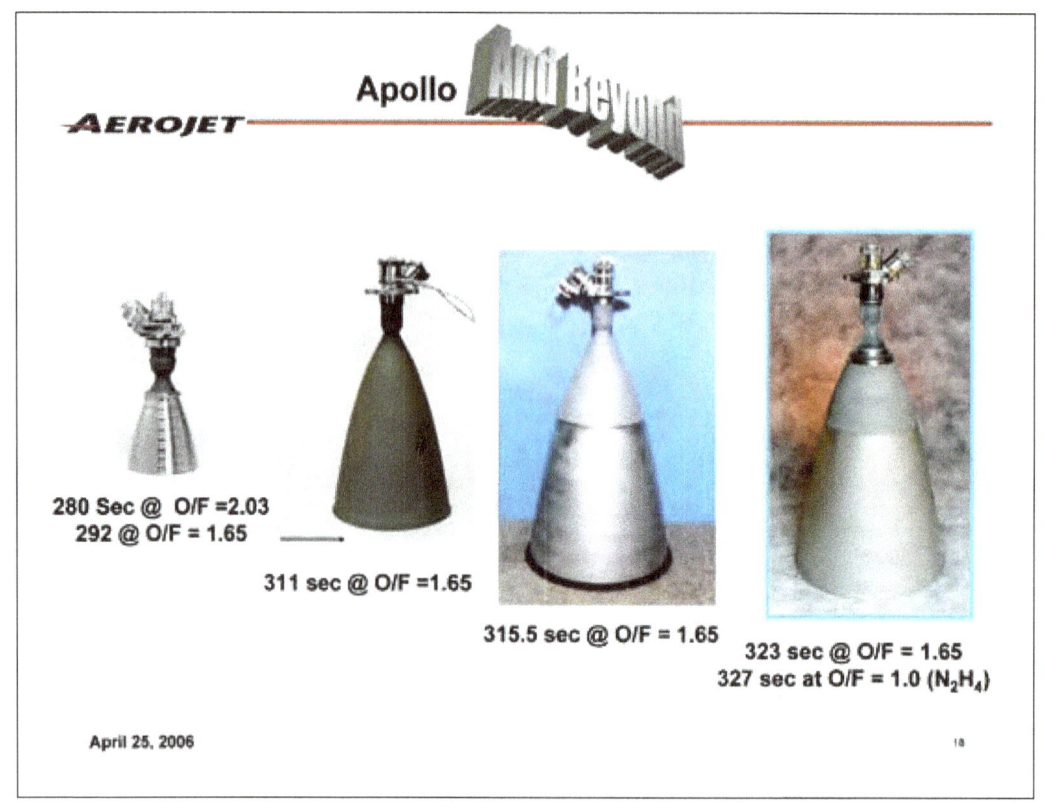

Appendix F

Tim Harmon's SE-7 & SE-8 Presentation Viewgraphs

Apollo Rocket Propulsion Development 133

Appendix F

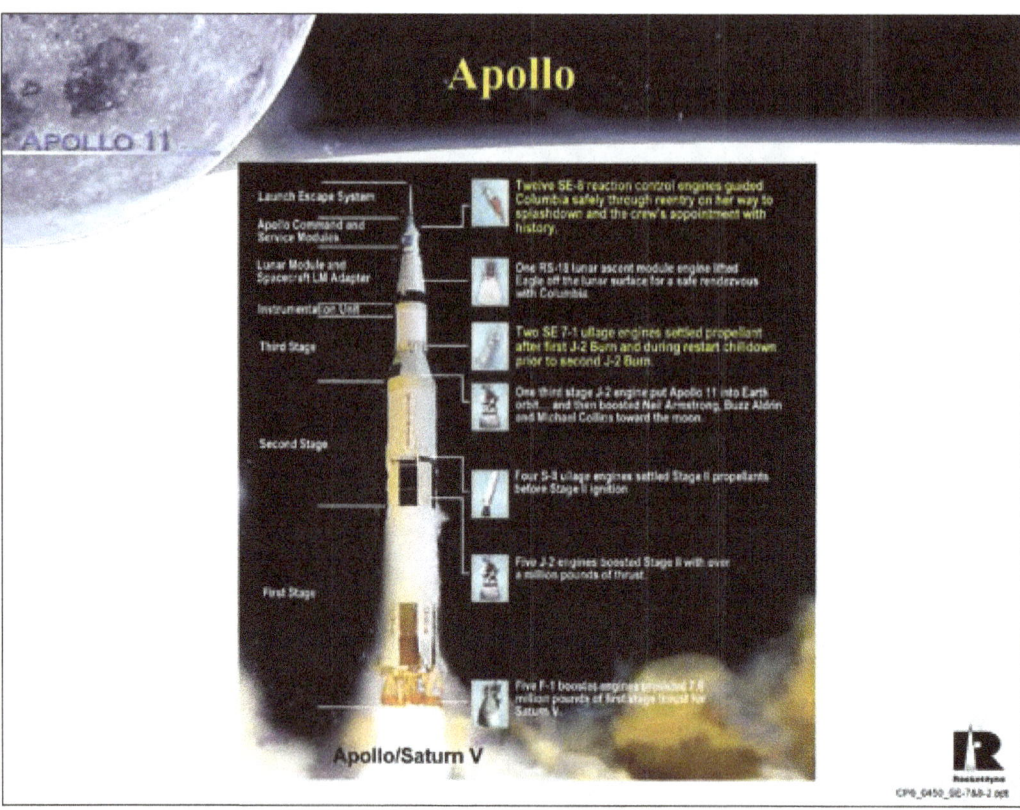

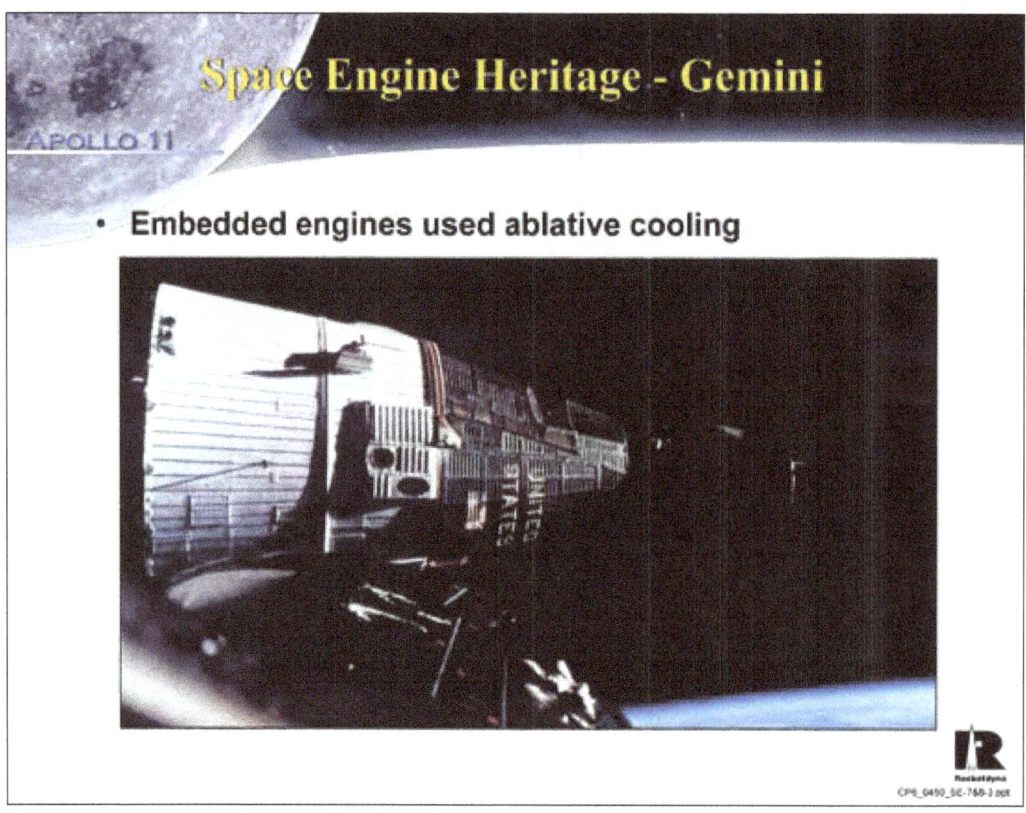

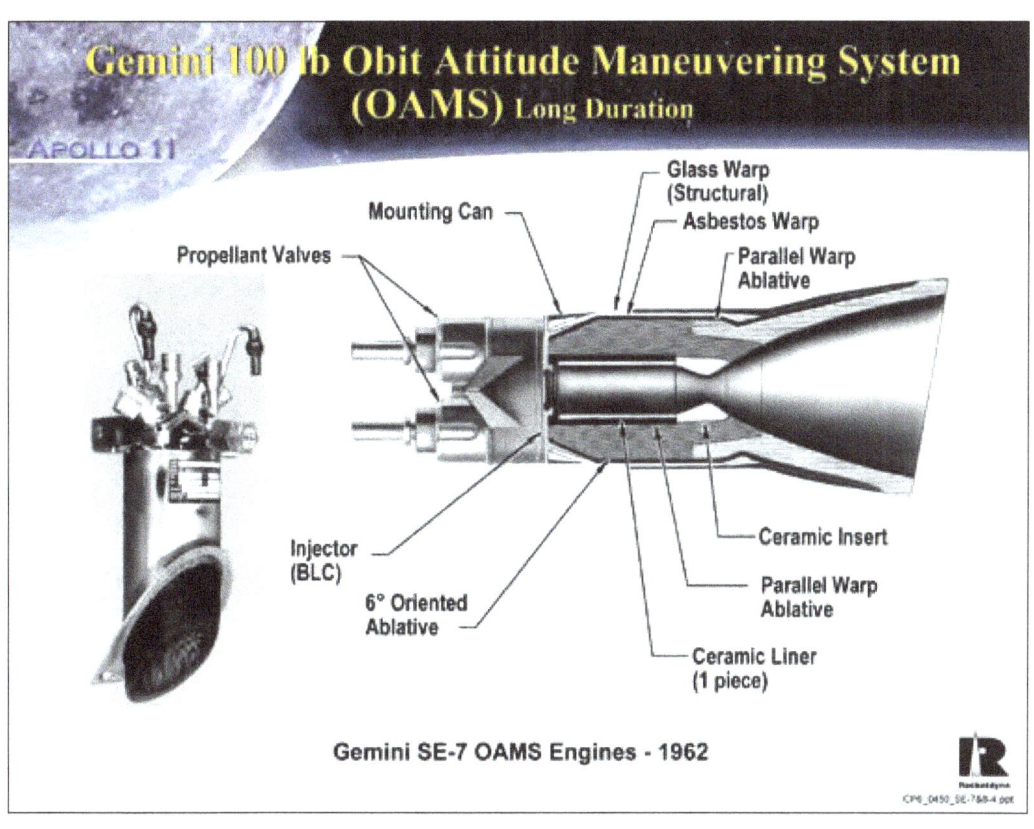

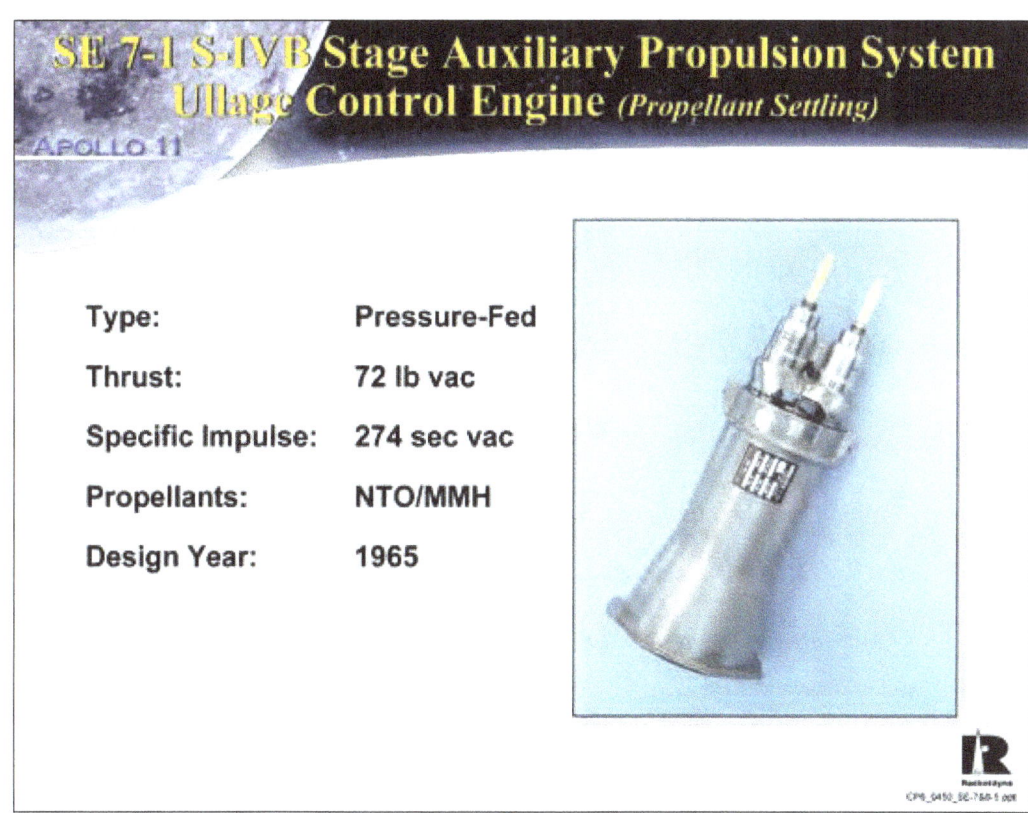

Appendix F

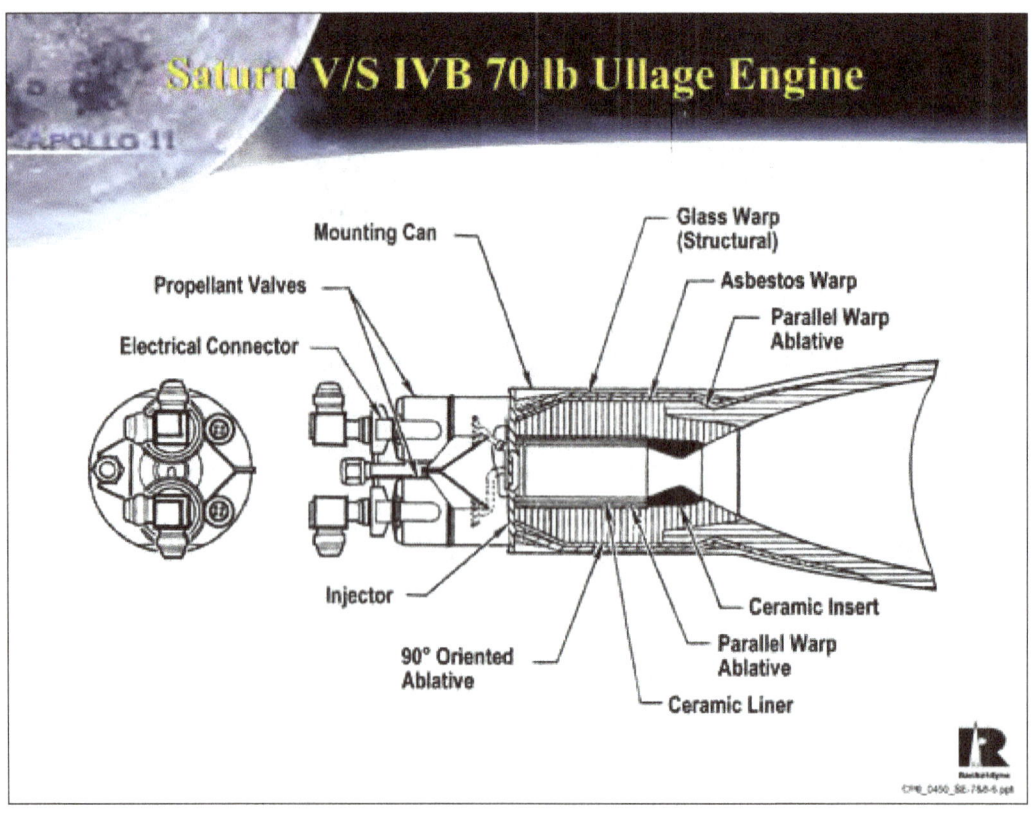

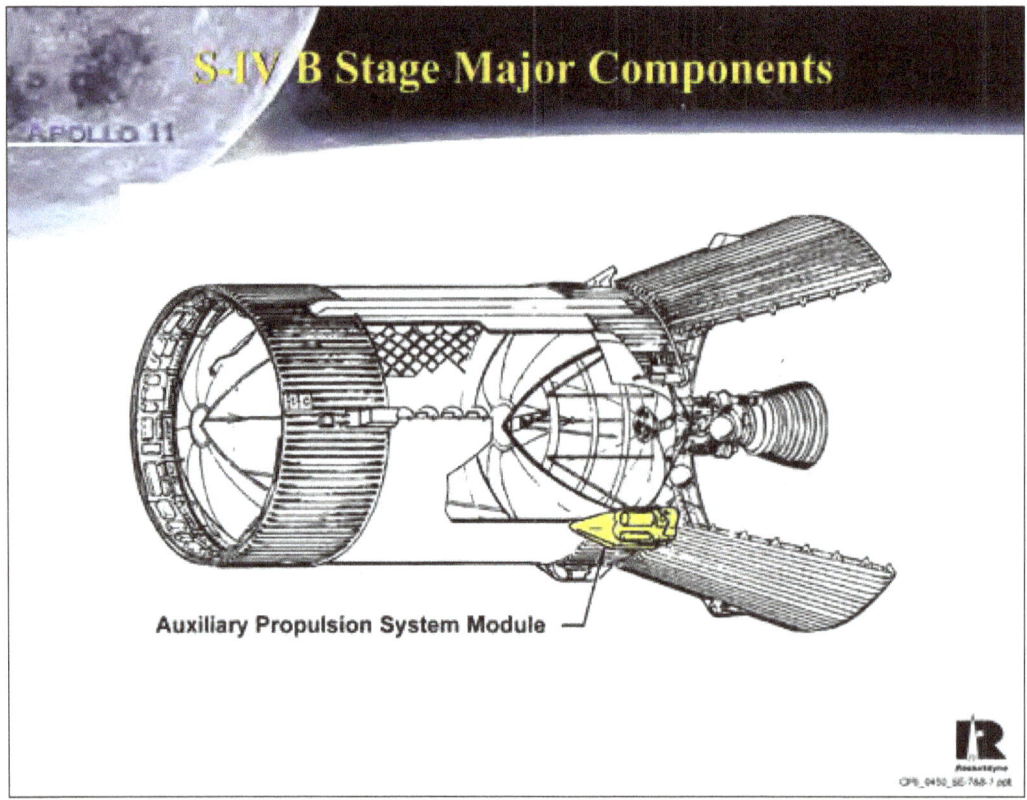

SE 7-1 Lessons Learned

- Engine proven in Space Environment
- No development issues
- No operational issues
- Cost effective application

Off the shelf applications work

SE 8 – Command Module Re-entry Control Engine

- Maintain heat shield orientation
  - 4 times the heat load over Gemini
- "Steer" the module for re-entry into atmosphere
  - Gemini re-entry speed   17,000 mph
  - Apollo re-entry speed   24,000 mph
- Two redundant systems of 6 engine each

Appendix F

SE-8 Command Module Reaction Control System
APOLLO 11

- Similar off-the-shelf approach
- Apollo Command Module reentry speed faster, heat shield thicker
- Engine "Improved"
  - ZrC throat insert
  - Pre-charred sleeve to eliminate ceramic liner
  - Single 93 lb thrust engine design, variable length nozzle extensions

SE-8 Apollo Command Module Reaction Control
APOLLO 11

Type:	Pressure-Fed
Thrust:	93 lb vac
Specific Impulse:	274 sec vac
Propellants:	NTO/MMH
Design Year:	1964

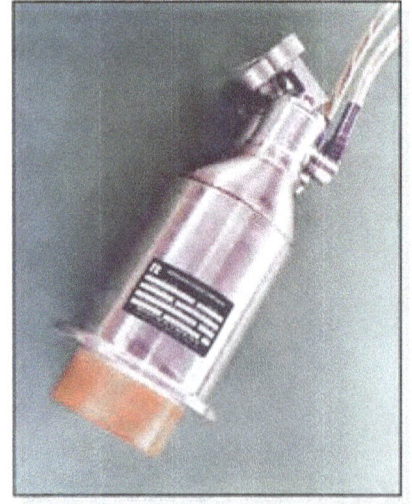

SE 8 Engine Post Mission Duty Cycle Testing
APOLLO 11

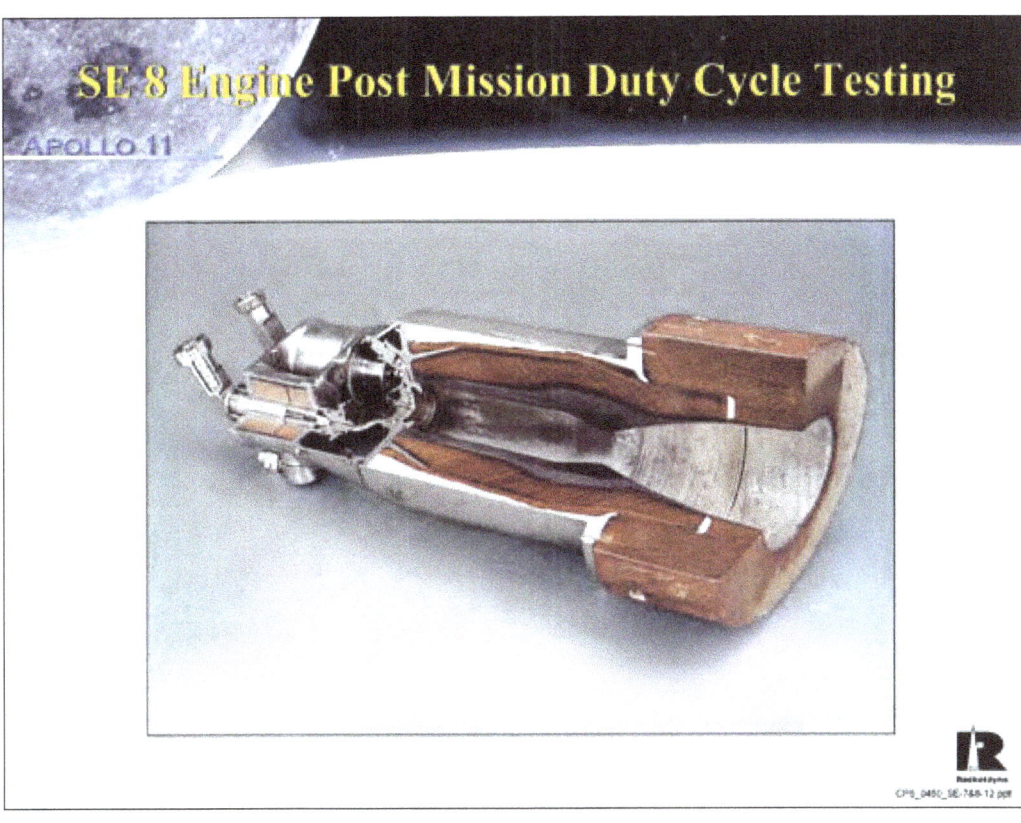

Each Injector Hot Fire Certified
APOLLO 11

- Issue – Some engines experienced uneven material ablation & throat insert erosion

- 12 Engines per vehicle…12 injector hot fire calibration cycles

- Subjective (visual) appearance required for approval

- Devised a tactile aid for inspectors to approve/disapprove throat wear

Appendix F

Component Test Lab IV, Santa Su

- Santa Su site of early B-movie westerns
- Two shift test operations
- Intense test slot competition
- Major data reduction effort

SE 8 Engine Test

- Test constraints
  - 15 minute altitude capability
  - Multiple data points desired
  - DIGR's (direct inking graphic recorder) instrumentation
  - Slide rule "calculators"

Tim Harmon's SE-7 & SE-8 Presentation Viewgraphs

SE-8 Lessons Learned

- Extensive test data & reports
- Devised 1 page test report template
    - Simplified report process and maintained report consistency
- Development issues resolved through extensive test program

Anticipate issues with off-the-shelf engines applied to new applications

Apollo Rocket Propulsion Development

Appendix G

Clay Boyce's Presentation Viewgraphs

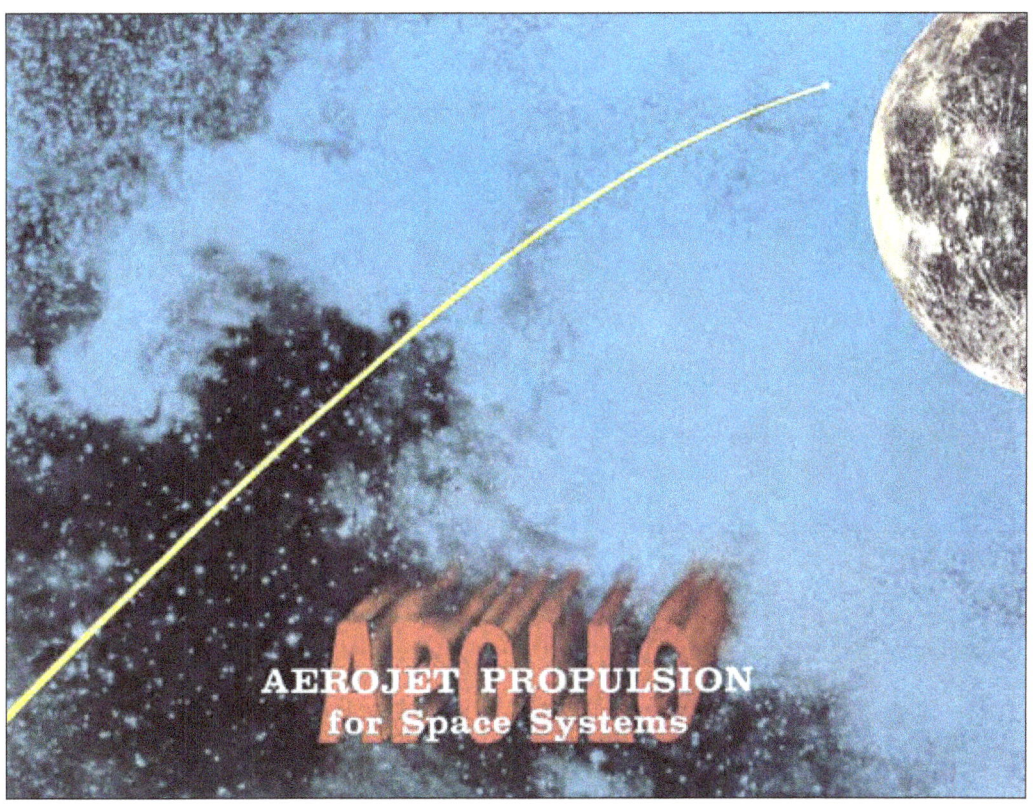

Appendix G

**BRIEF CHRONOLOGY OF EVENTS
DURING DEVELOPMENT
OF THE

APOLLO SERVICE MODULE
ROCKET ENGINE**

Clay Boyce

Aerojet (Retired)

Sacramento, CA

Presented at

Stennis Space Center

April 25, 2006

APOLLO SPS ENGINE

- Proposal Task Start July 1960

- Contract Start April 1962

- Pre LEM Decision Concept

APOLLO SPS ENGINE

AEROJET

- Technology Base
    - Saint Program
    - Apollo Subscale

APOLLO SPS ENGINE

AEROJET

Thrust	20,000 lb
Propellants	NTO/A-50
Chamber Press	100 psia
Inlet Press	165 psia
Exp Ratio	62.5 to 1
Duration	750 sec
Restarts	50

Appendix G

APOLLO SPS ENGINE
AEROJET

- Flight Firsts
    - Ablative Thrust Chamber
    - Throat Gimbal
    - Columbium Nozzle

APOLLO SPS ENGINE
AEROJET

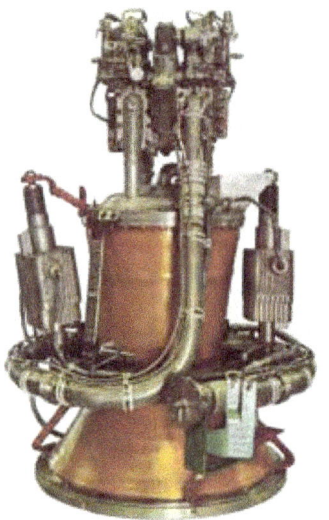

- **Final Powerhead Assembly**
    - 1st Firing Surprise
    - SPS Module Interface Surprise

APOLLO SPS ENGINE

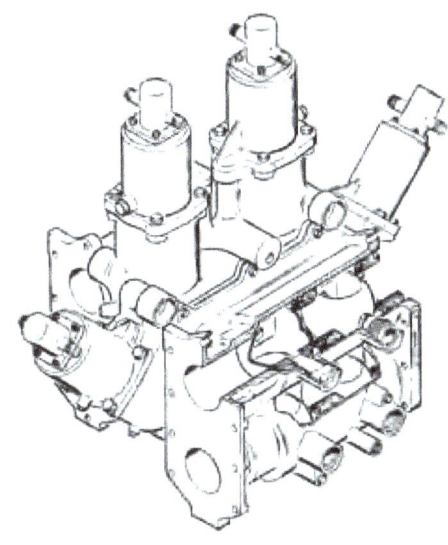

- Propellant Valve Assembly
  - NC Machining
  - Actuation

APOLLO SPS ENGINE

- Injector Assembly
  - Aluminum Brazing
  - Popping
  - Dynamic Stability

Appendix G

APOLLO SPS ENGINE
AEROJET

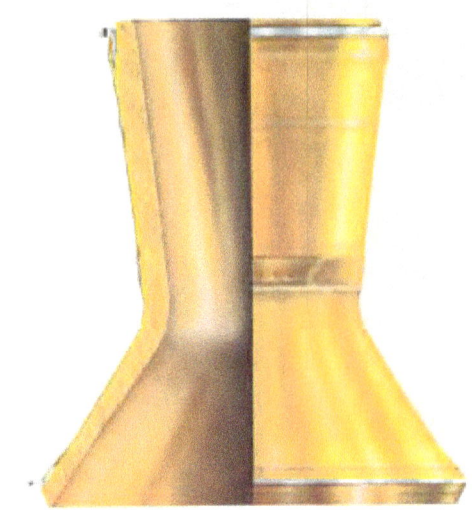

- Thrust Chamber Assembly
    - Ablative Material
    - Flanges

APOLLO SPS ENGINE
AEROJET

- Gimbal Actuator Assembly
    - Contract Events
    - Design Events

APOLLO SPS ENGINE

AEROJET

- Nozzle Assembly

    - Gore Fabrication

    - 1st Hot Fire

    - Columbium

APOLLO SPS ENGINE

AEROJET

- **Firing History**

	No. Missions	No. Starts	Total Firing Time (sec)
Apollo Unmanned	4	9	1,232
Apollo Manned	11	64	5,060
Skylab Manned	3	20	115
Apollo/Soyuz	1	9	18
Total	19	102	6,425

First SPS flight was February 26, 1966
Longest single firing on a mission was 445 sec. (Apollo 6, Apr. 4, 1968)
Shortest firing on a mission was 0.5 sec. (Apollo 7, Oct. 11, 1968)
Most firings for an engine on a mission was 8 (Apollo 7, 9, and 15)

Appendix G

Appendix H

Jerry Elverum's Presentation Viewgraphs

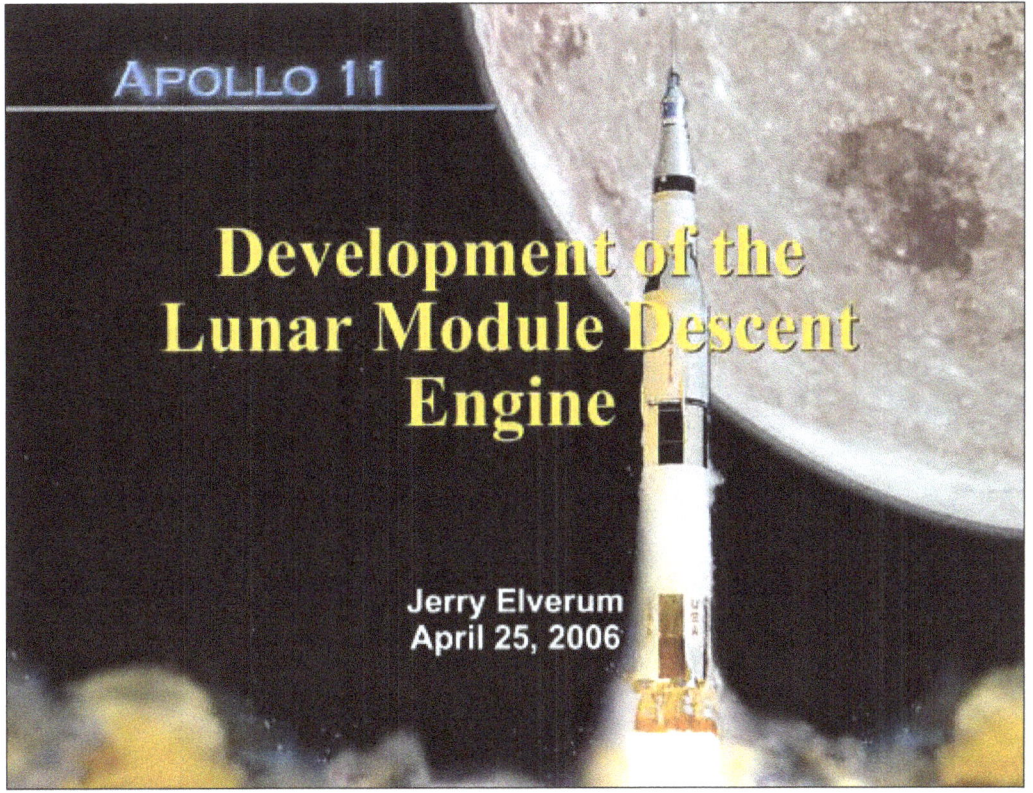

Appendix H

FIGURE 2
VARIABLE THRUST ROCKET ENGINE APPLICATION AREAS

SYSTEM	MANEUVER	PURPOSE	VEHICLE TYPE
SPACE VEHICLE REACTION CONTROL	EXTRATERRESTRIAL SOFT LANDING	ATTITUDE & VELOCITY CONTROL DURING DESCENT AND HOVER PHASE	SURVEYOR PROSPECTOR
	RENDEZVOUS AND DOCKING	AXIAL & LATERAL VELOCITY CORRECTIONS FOR PRECISE ORBIT TRANSFER	SAINT SATURN STAGE
	REENTRY	ATTITUDE CONTROL REACTION FORCES FOR ORIENTATION & ATTITUDE DAMPING	APOLLO
WEAPON SYSTEM PROPULSION	PROPORTIONAL NAVIGATION	WEAPON ACCELERATION PROPORTIONAL TO TARGET-INTERCEPTOR SPECIAL RELATIONSHIP	BAMBI
	VACUUM TRAJECTORY	THRUST EQUAL TO DRAG AT VARIOUS ALTITUDES	MISSILE B

METHODS OF ACHIEVING THRUST LEVEL MODULATION

1. FIXED AREA INJECTOR WITH SEPARATE FLOW CONTROL VALVE

2. VARIABLE AREA INJECTOR

3. VARIABLE AREA INJECTOR WITH SEPARATE FLOW CONTROL VALVE

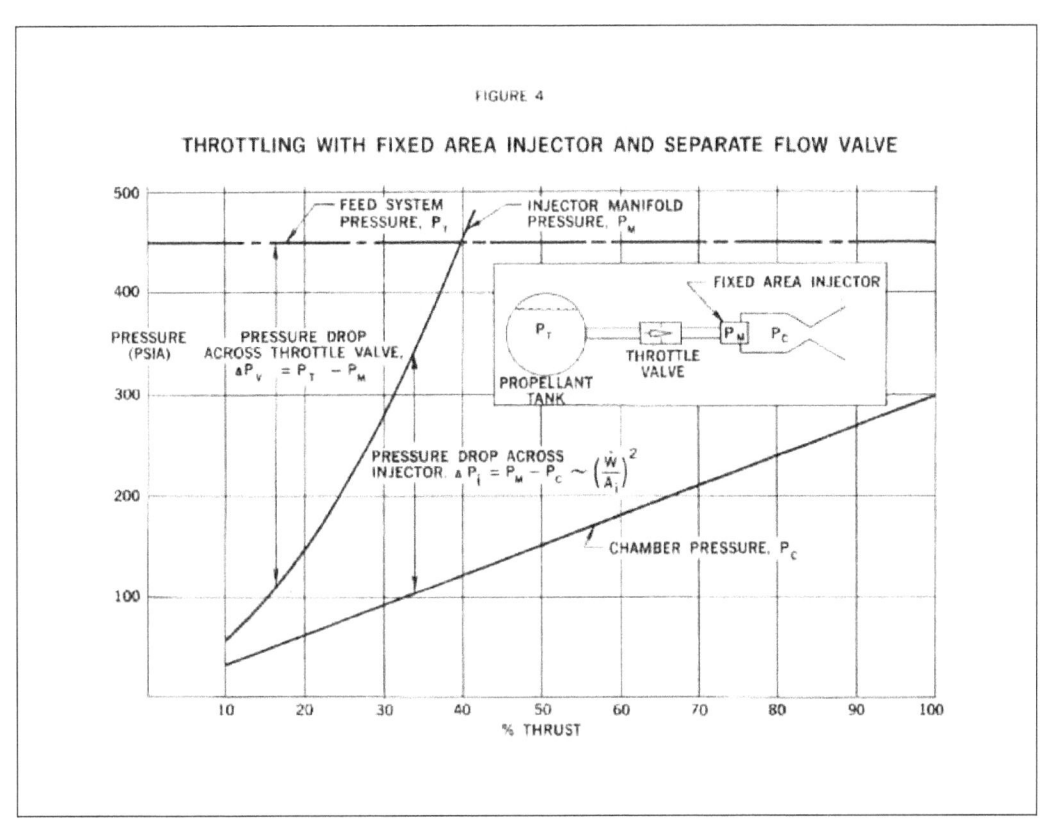
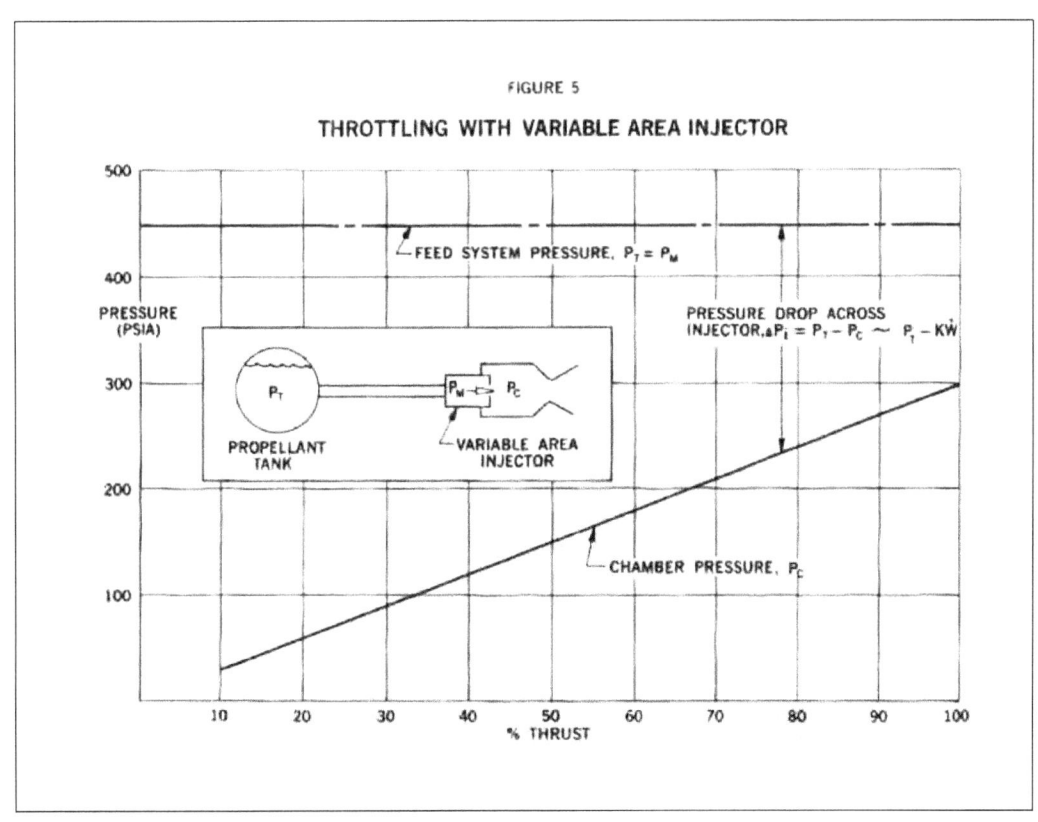

Appendix H

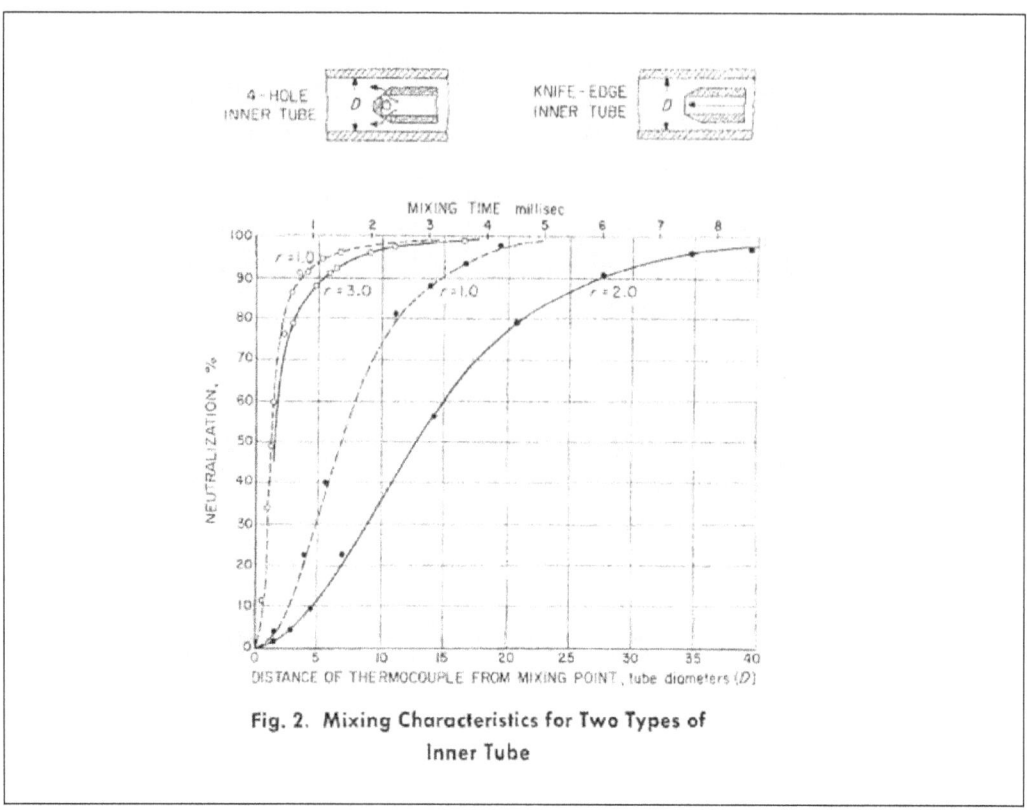

Fig. 2. Mixing Characteristics for Two Types of Inner Tube

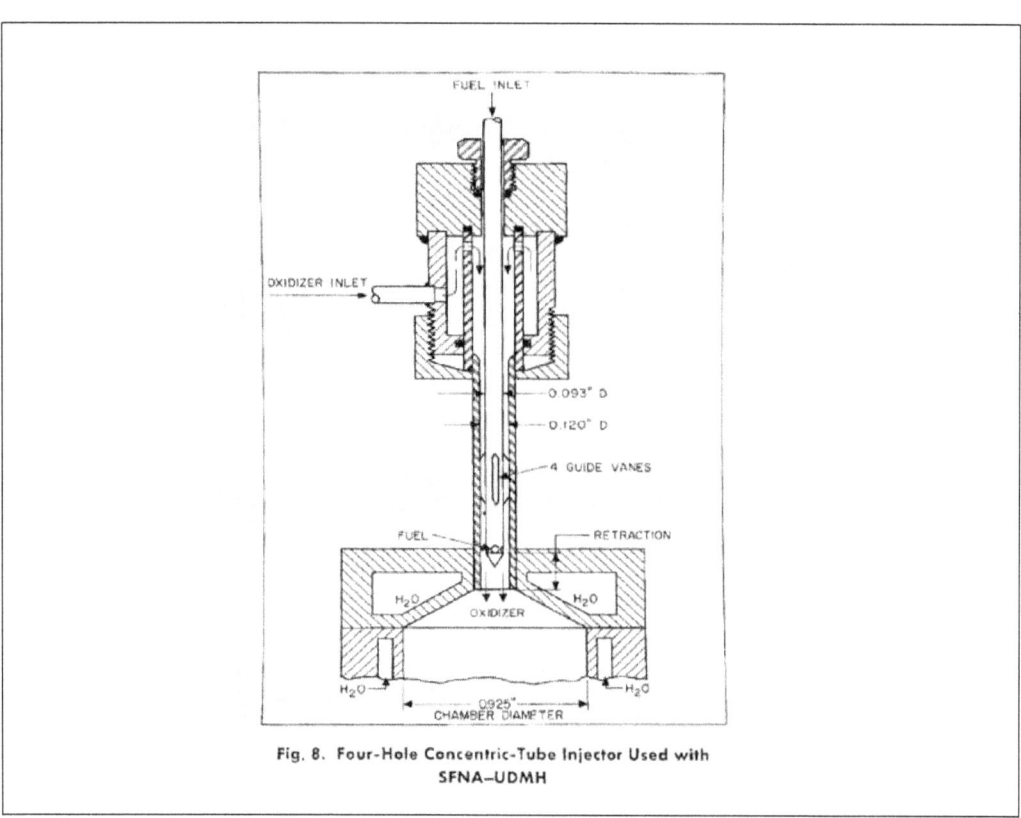

Fig. 8. Four-Hole Concentric-Tube Injector Used with SFNA–UDMH

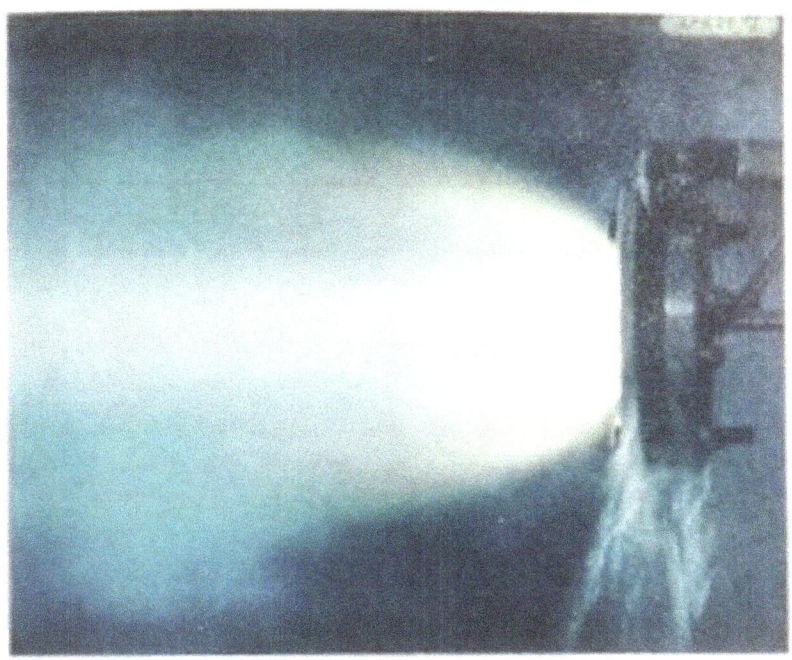

Plate 26. Concentric-Tube Injector of Plate 25 with 0.15-in.-long Chamber

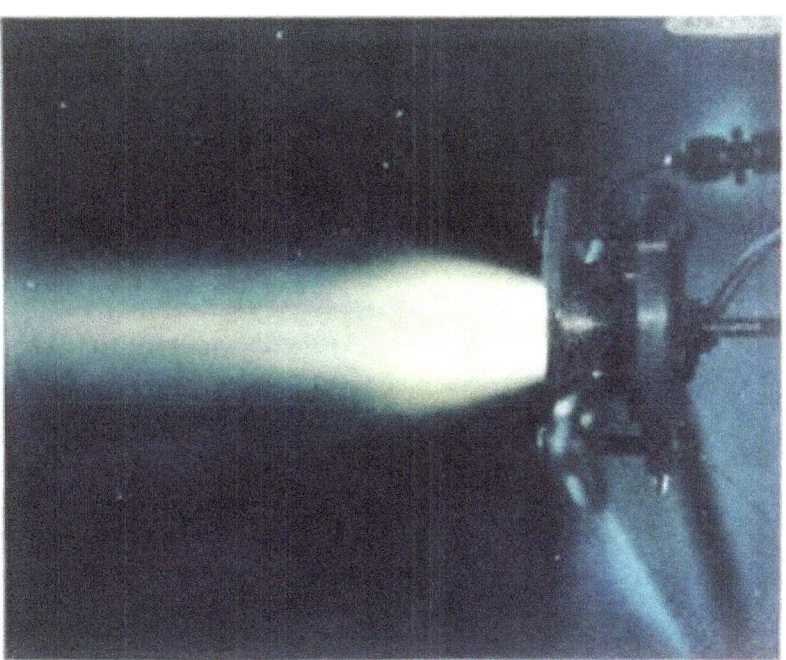

Plate 27. Concentric-Tube Injector of Plate 25 with 1.0-in.-long Chamber

Appendix H

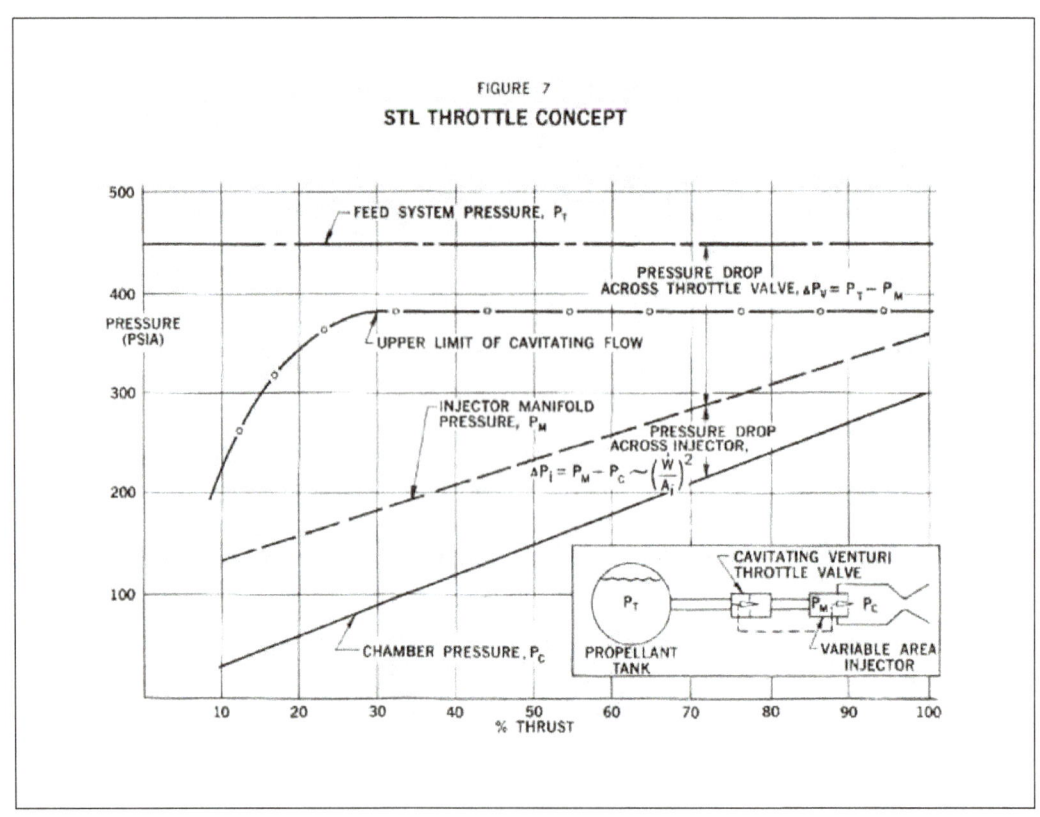

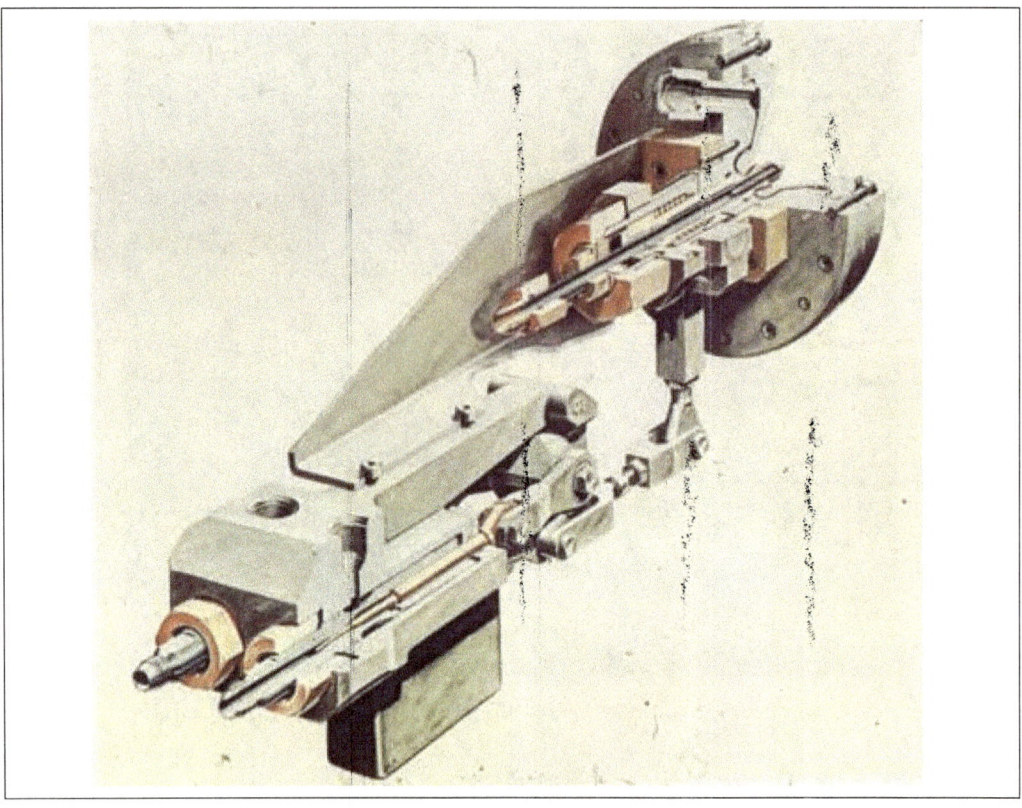

Design of the Pintle Mechanism.

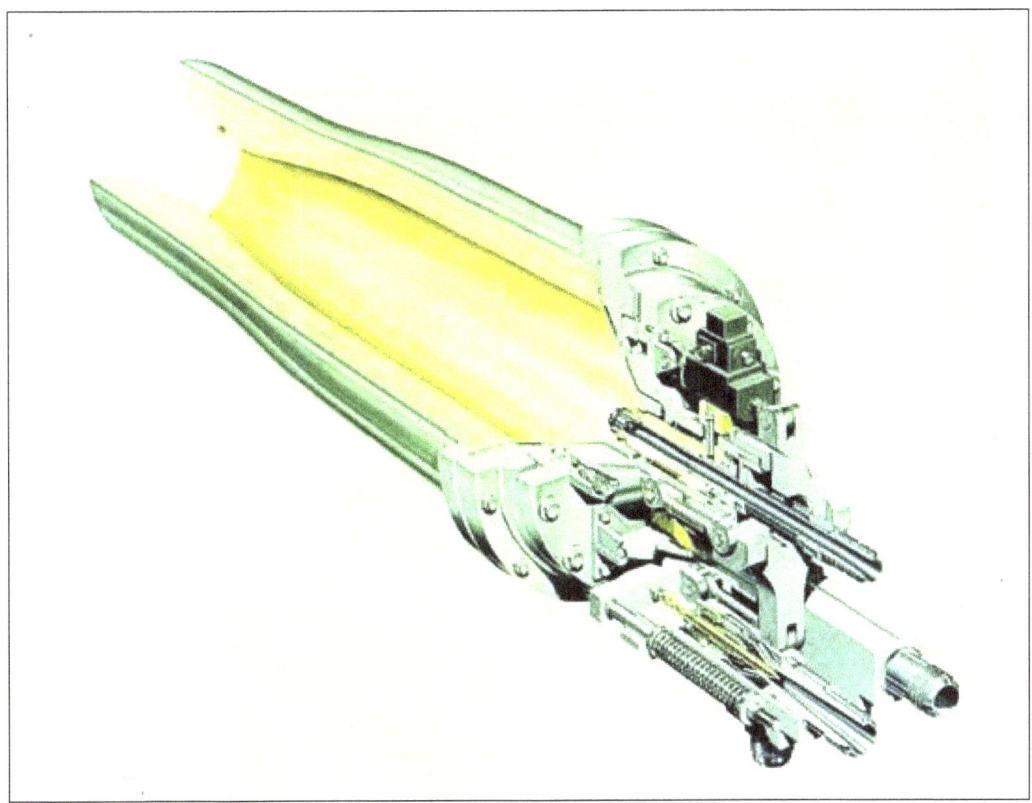

Pintle injection thrust chamber illustration.

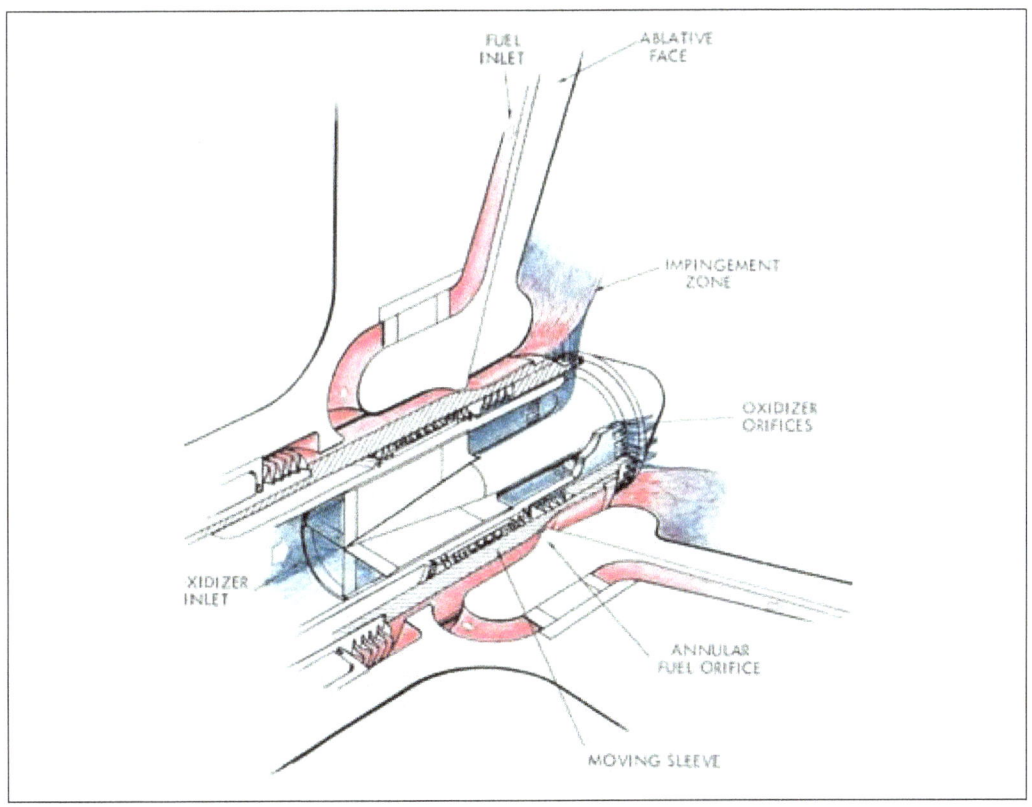

Appendix H

Hot fire test facility

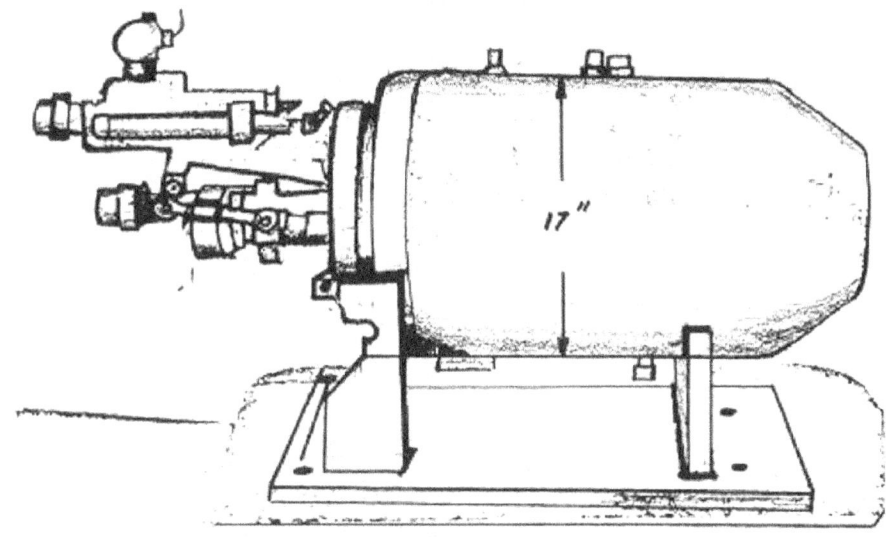

FIGURE 14. COMBUSTION STABILITY TEST CHAMBER
"IRON PIG"

5000 LB THRUST THROTTLING ENGINE

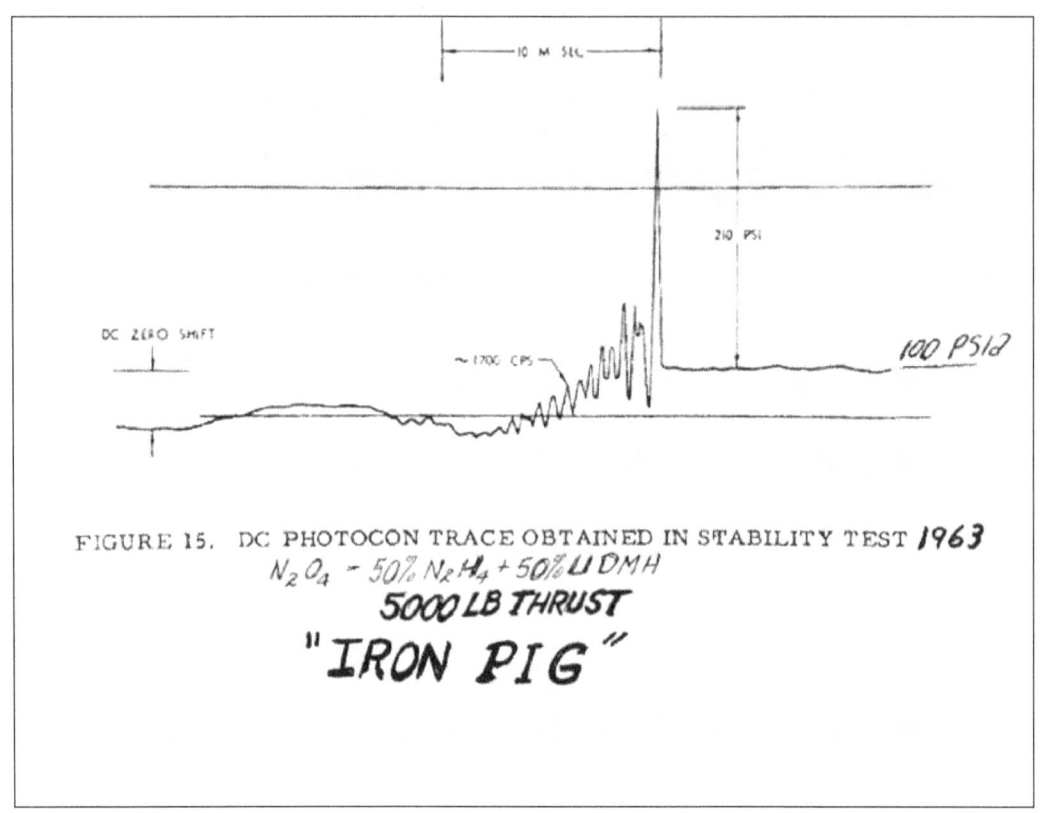

FIGURE 15. DC PHOTOCON TRACE OBTAINED IN STABILITY TEST 1963
N_2O_4 – 50% N_2H_4 + 50% UDMH
5000 LB THRUST
"IRON PIG"

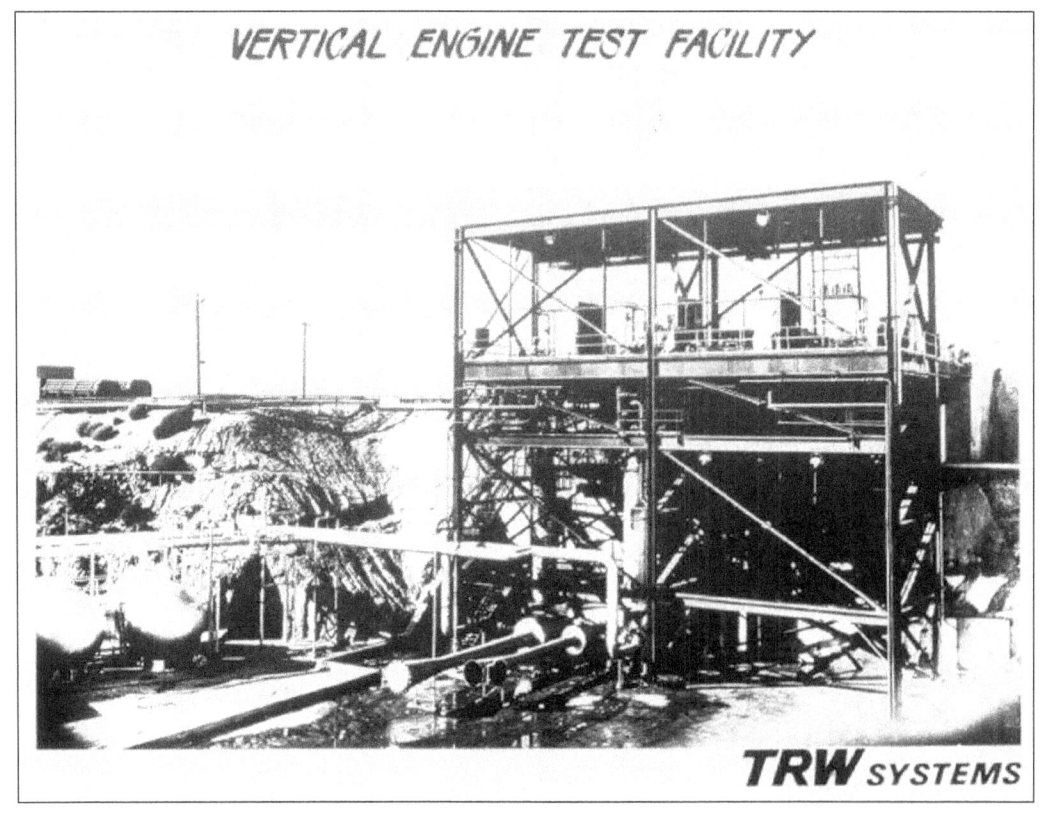

Jerry Elverum's Presentation Viewgraphs

Altitude simulation test stand

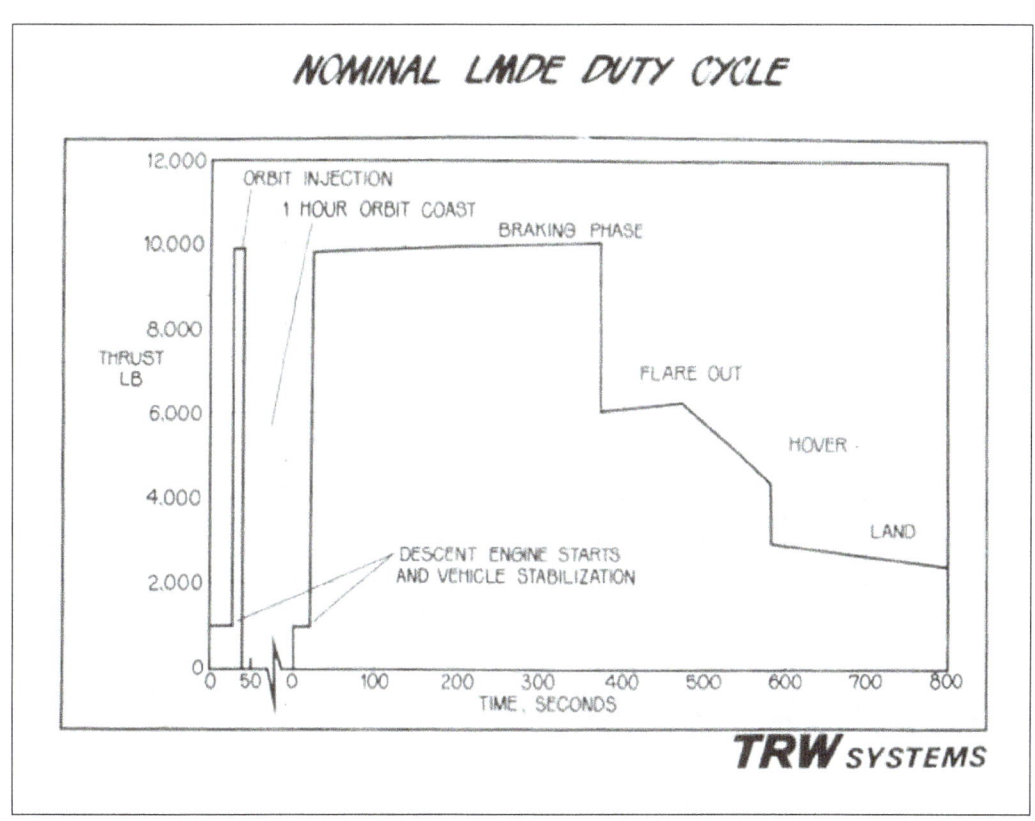

Apollo Rocket Propulsion Development 161

Appendix H

```
LEM DESCENT ENGINE DESIGN AND METHOD
OF OPERATION TO MEET LANDING MISSION

CO-AXIAL CENTRAL INJECTOR ELEMENT WITH SINGLE MOVING
   SLEEVE TO VARY OXIDIZER AND FUEL OR'FICES.

SEPERATE FLOW CONTROL VALVE, NON-CAVITATING AT
   FULL THROTTLE POSITION, CAVITATING FROM 60 TO 10%
   THROTTLE POSITIONS.

ENGINE DESIGNED TO PERFORM HIGH THRUST PORTION OF
   THE DUTY CYCLE AS A CALIBRATED ENGINE AT FULL
   THROTTLE POSITION.

DURING THROTTLE PORTION OF DUTY CYCLE AT 60 TO 10%
   THRUST, PROPELLANT FLOW RATES FULLY CONTROLLED
   BY CAVITATING VALVES.

THRUST CHAMBER DESIGN: TITANIUM CASE LINED WITH
   SILICA-PHENOLIC ABLATIVE MATERIAL OUT OF NOZZLE
   EXPANSION RATIO OF 16:1; COLUMBIUM TO 48:1.
```

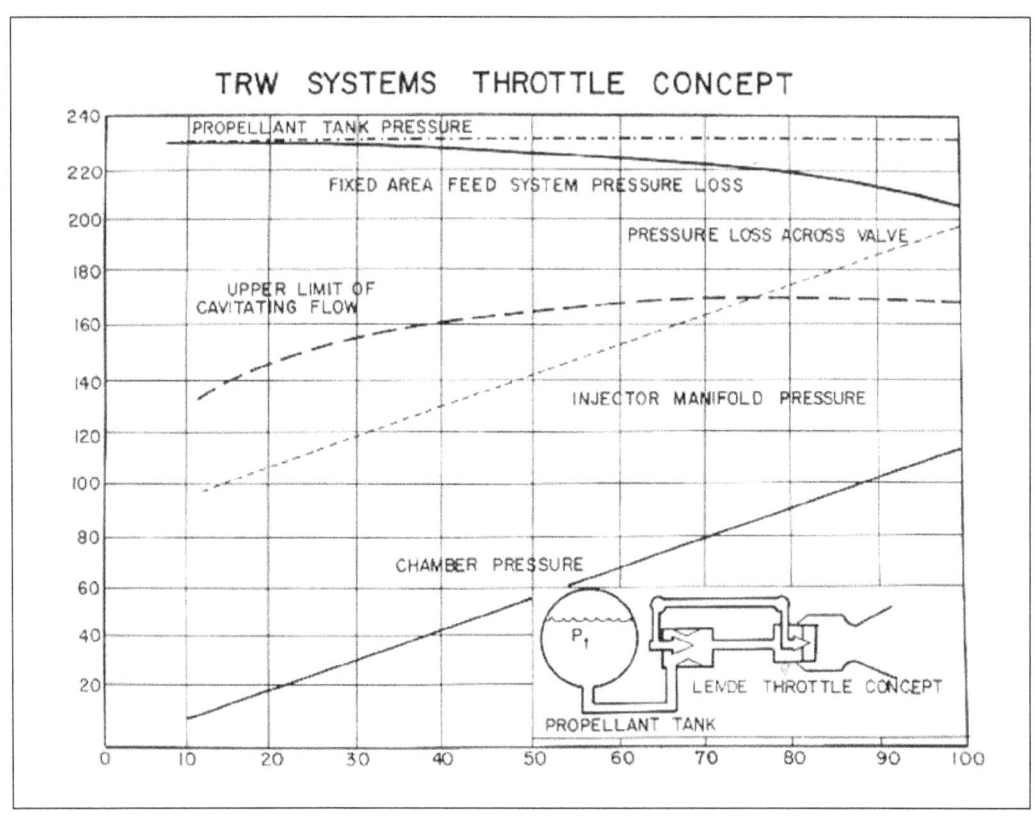

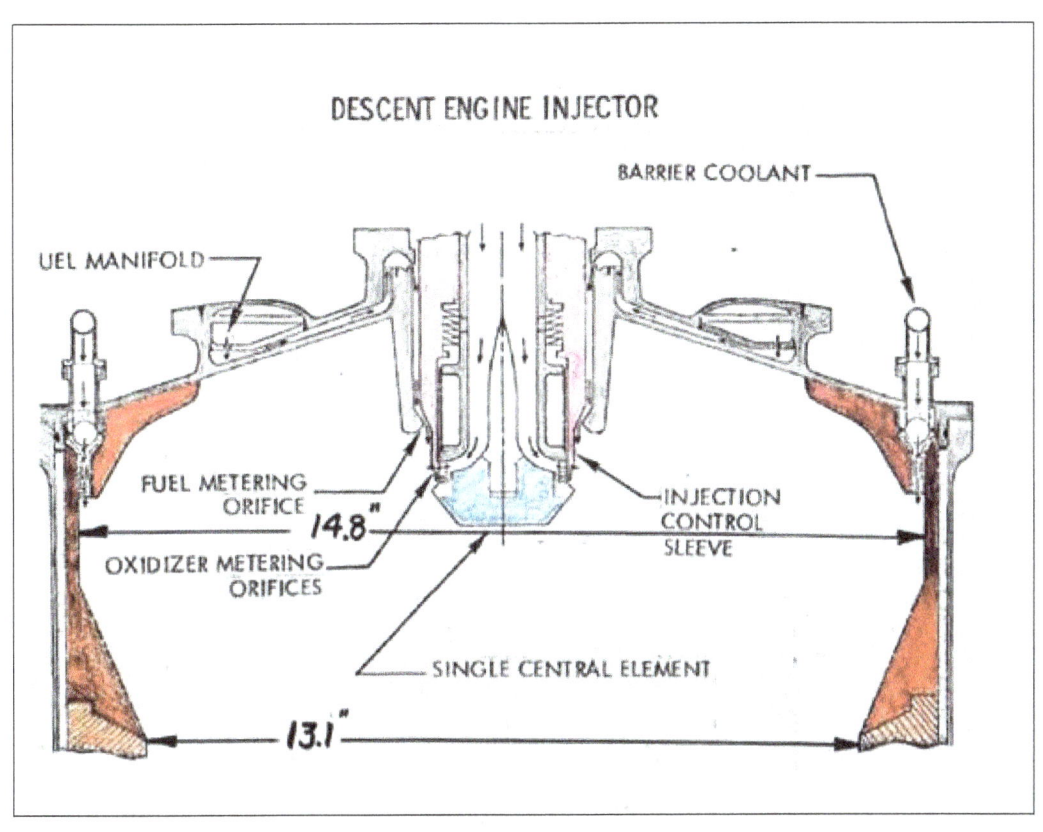
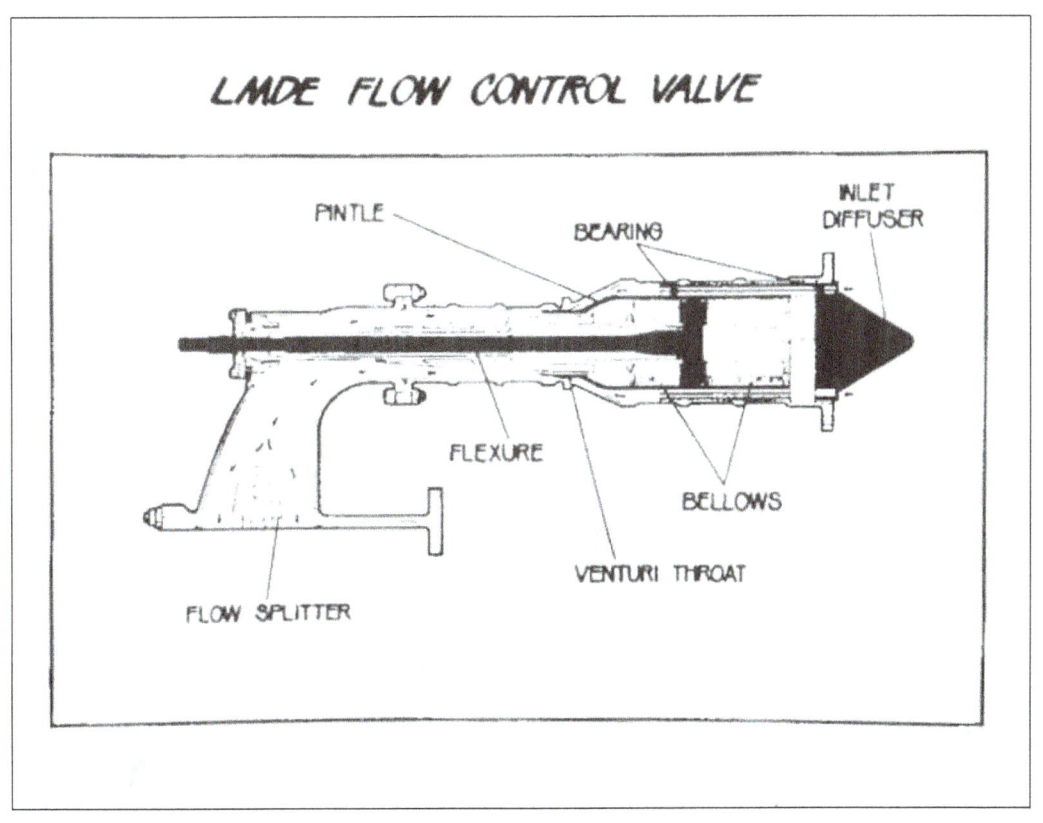

Appendix H

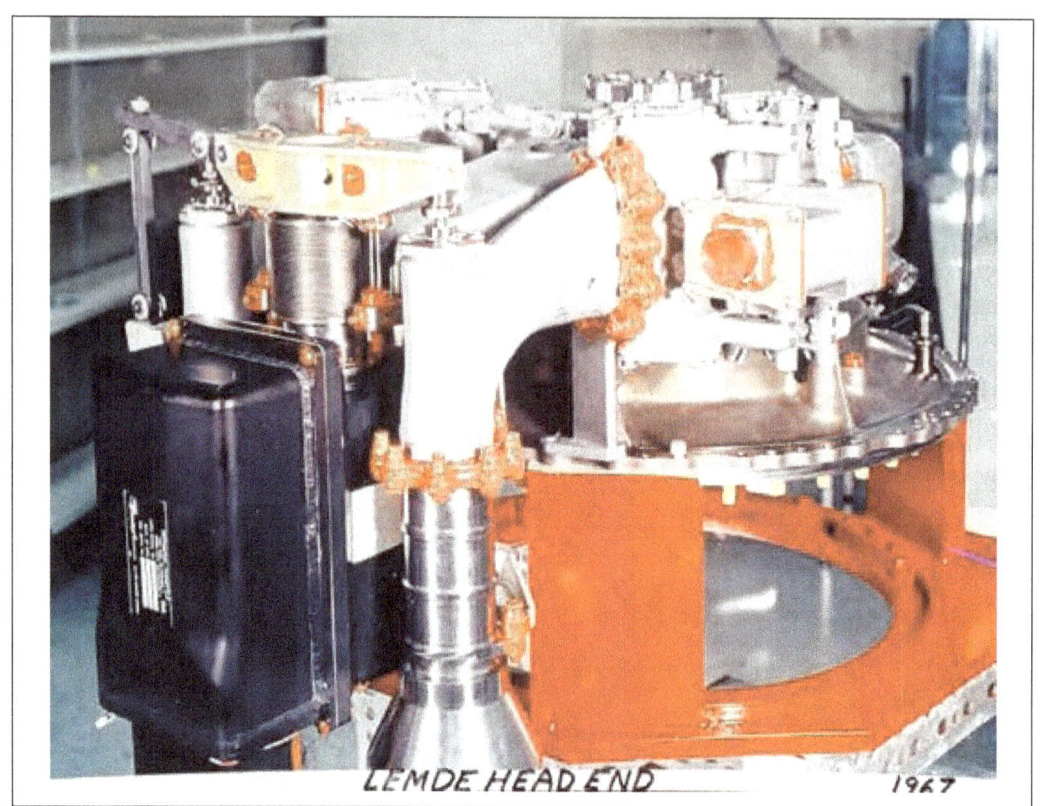

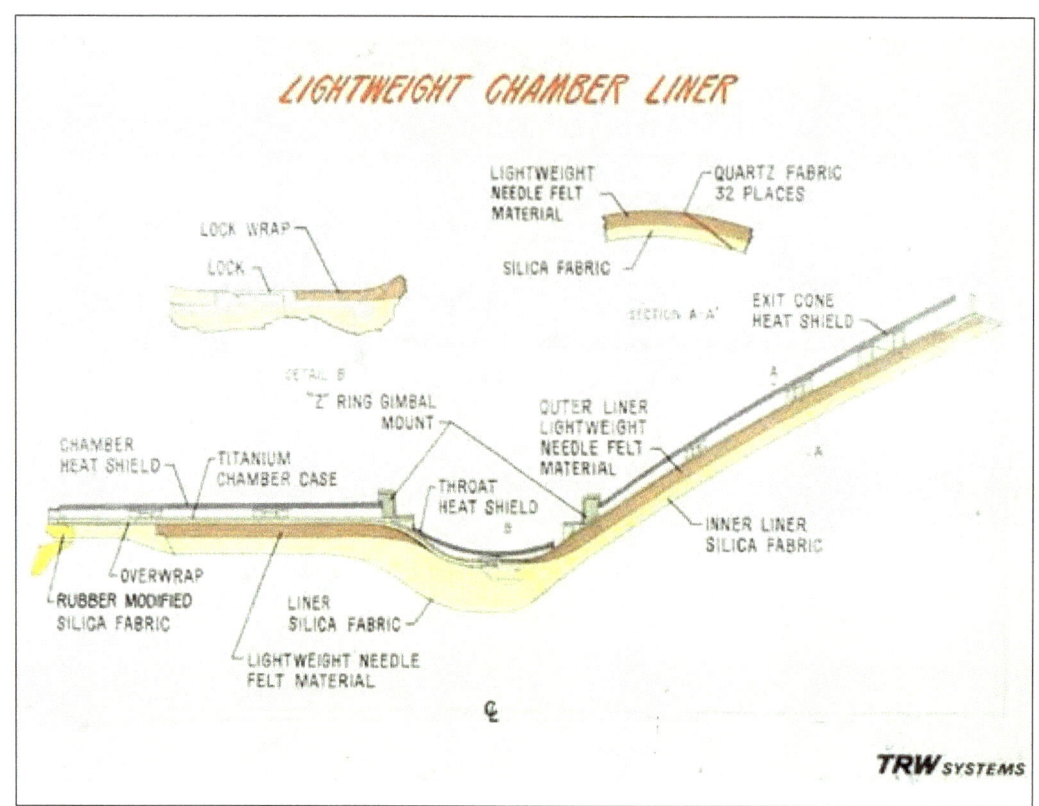

Jerry Elverum's Presentation Viewgraphs

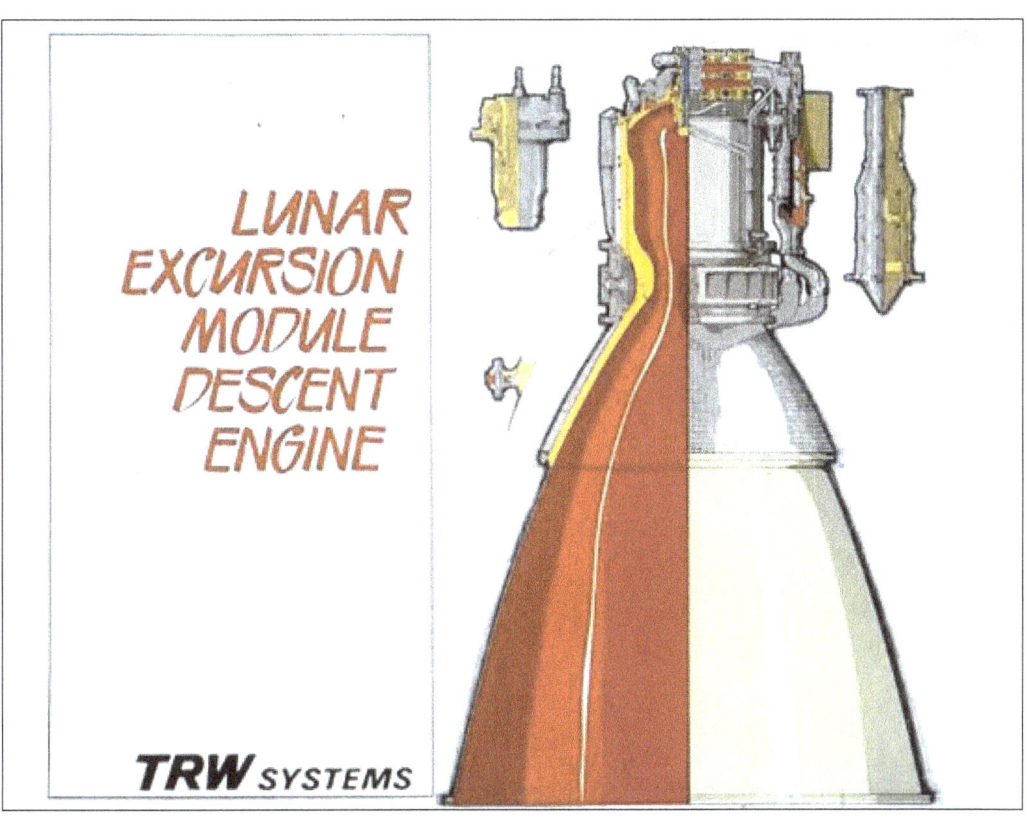

COMBUSTION STABILITY

- 31 "BOMB" TESTS CONDUCTED WITH 5-50 GRAIN CHARGES.
- INDUCED SPIKE >175% P_C, ALL TESTS.
- TESTS AT 10%, 25%, 50%, AND 100% THRUST.
- TYPICAL RECOVERY:

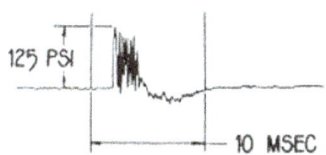

RECOVERY TIME <40 MSEC. ALL TESTS
- HIGH FREQUENCY INSTABILITY HAS NEVER BEEN SUSTAINED

TRW SYSTEMS

Apollo Rocket Propulsion Development 165

Appendix H

Major Engine Development History

Engine	Thrust Level	Instability
F1	1.5×10^6	Yes
H-1	188K	Yes
J-2	205K	Yes
Thor	150K	Yes
Titan II	215K	Yes
Titan II	100K	Yes
LMDE	10.5K	None
LM Ascent (Bell)	3.5K	Yes
Apollo SM	21.9K	Yes
Transtage	8K	Yes

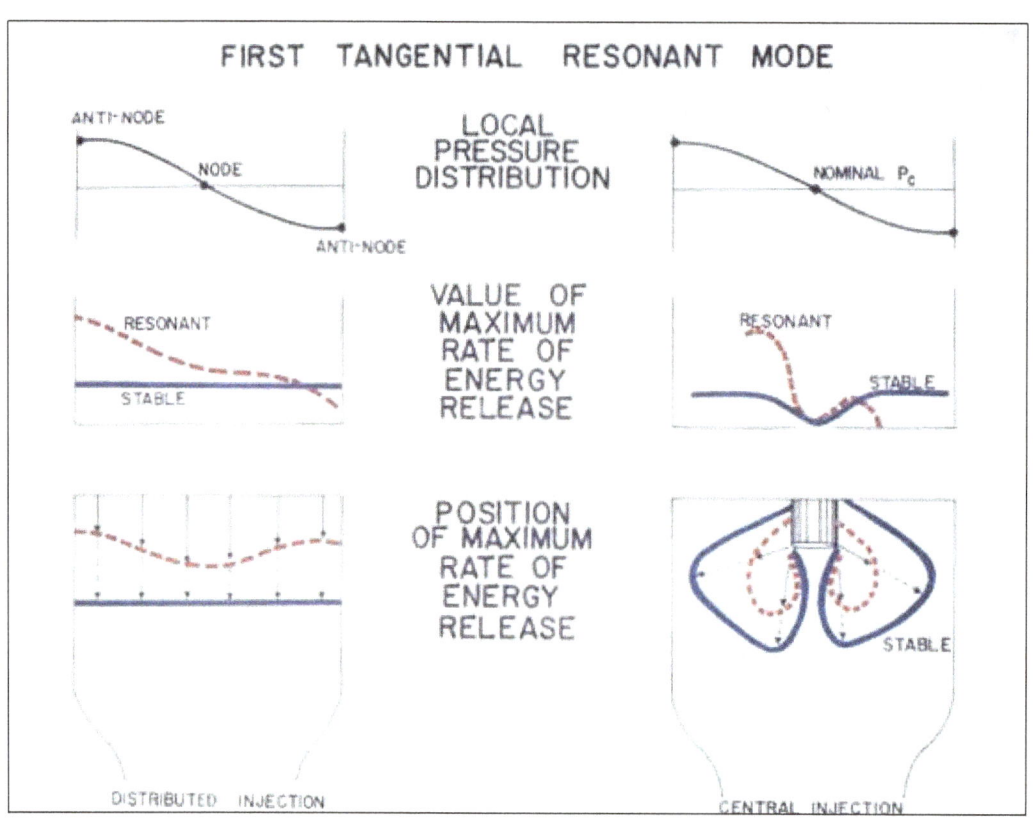

LMDE TEST SUMMARY
(THRU MARCH. 1967)

	NUMBER OF NEW BUILDS	NUMBER OF STARTS	FIRING DURATION (SECONDS)
INJECTOR TESTS		1,755	70,191
THROTTLING HEAD END ASSEMBLIES **	26	816	55,463
ABLATIVE ENGINES:			
• SEA LEVEL ($\epsilon = 2:1$)	20	43	4,357
• HIGH ALTITUDE ($\epsilon = 48:1$)	27	195	18,834
TOTAL TEST EXPERIENCE		2,809	148,845 ***

TRW SYSTEMS

LMDE
PROGRAM SUMMARY

- ROCKETDYNE ENGINE PROGRAM STARTED — FEBRUARY 1963
- *TRW SYSTEMS* PROGRAM STARTED — JULY 1963
- FIRST ENGINE TEST IN HIGH ALTITUDE TEST STAND — AUGUST 1964
- FULL THROTTLING RANGE ENGINE TEST — NOVEMBER 1964
- *TRW SYSTEMS* SELECTION — JANUARY 1965
- INITIATION OF PHASE A QUALIFICATION — JULY 1966
- COMPLETION OF PHASE A QUALIFICATION — NOVEMBER 1966
- COMPLETION OF PHASE B QUALIFICATION SCHEDULED — JULY 1967
- 9 ENGINES DELIVERED TO CUSTOMER — MAY 1967

TRW

Appendix H

Jerry Elverum's Presentation Viewgraphs

Appendix H

Appendix I

Tim Harmon's Lunar Ascent Engine Presentation Viewgraphs

Appendix I

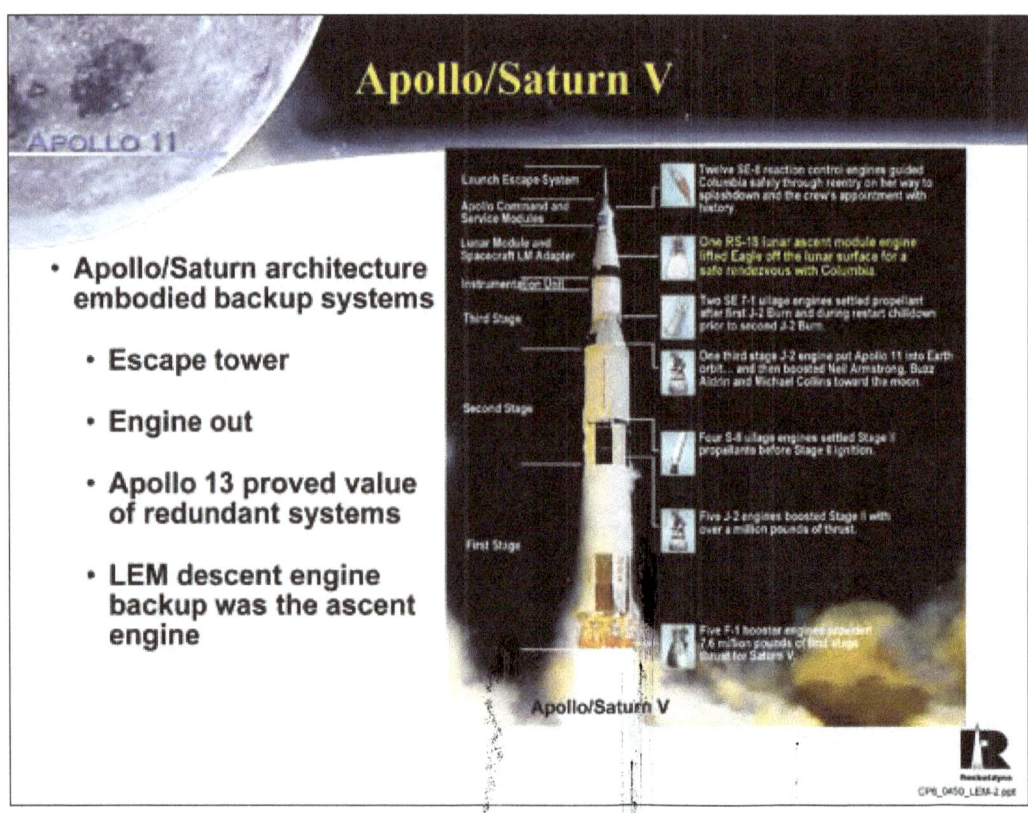

Tim Harmon's Lunar Ascent Engine Presentation Viewgraphs

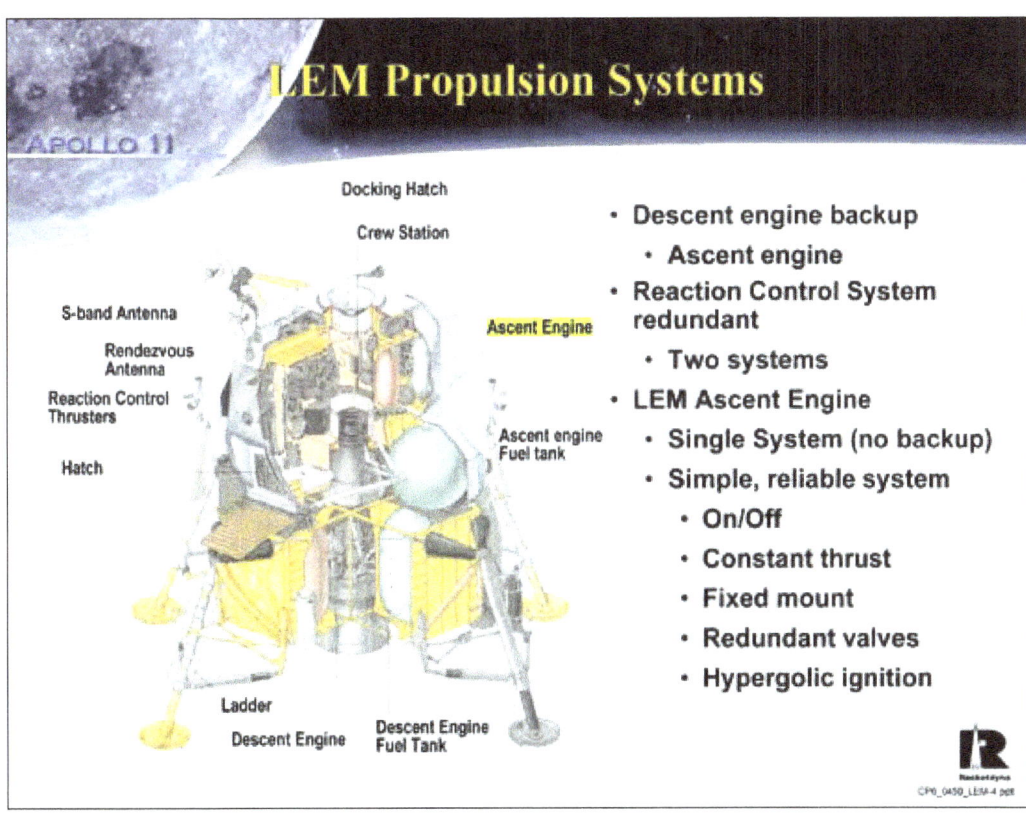

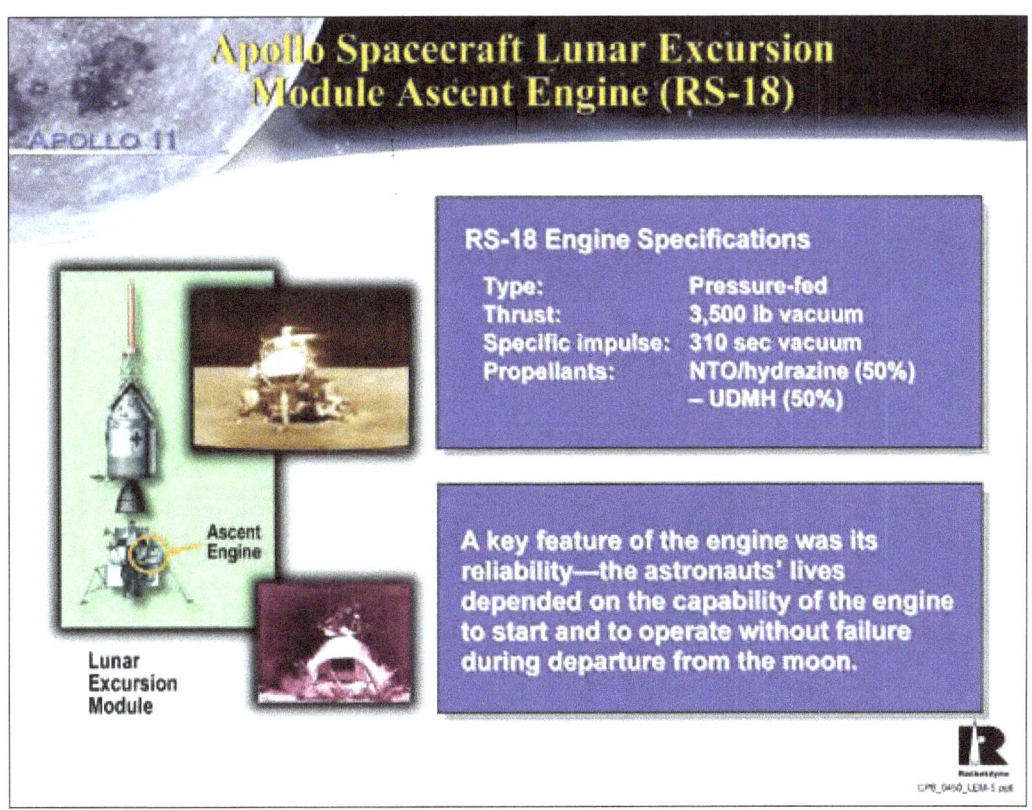

Apollo Rocket Propulsion Development

Appendix I

Ascent Engine Development Program

- Bell Aerospace began development in July of 1963
- Challenges with the injector
  - Thrust chamber erosion
  - Combustion Instability
- Problems continued & NASA awarded backup injector contract to Rocketdyne in August 1967
  - Solved injector erosion and instability challenges
- In September 1968 Rocketdyne was awarded the LEM engine contract by NASA

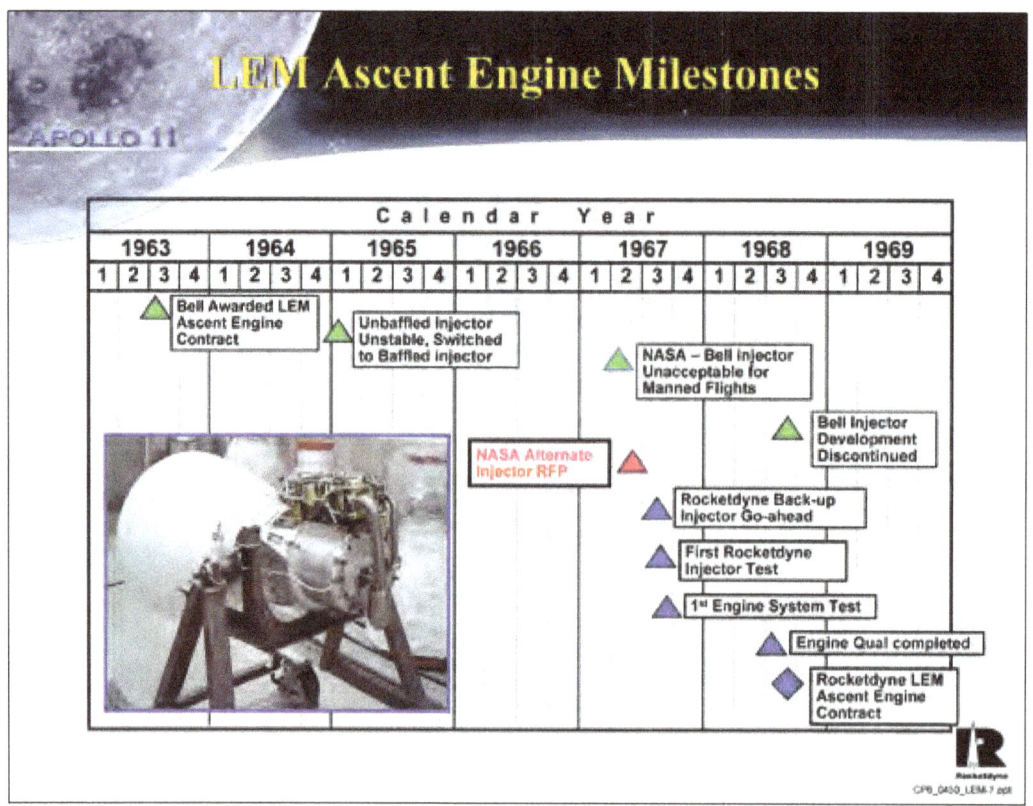

LEM Ascent Engine Milestones

Rocketdyne Development Program Milestones

APOLLO 11

- Program go-ahead — 8/3/1967
- 1st injector test — 9/7/1967
- 1st engine system altitude test — 9/20/1967
- 1st full duration test — 10/4/1967
- Engine system PDR completed — 11/1/1967
- Engine system CDR completed — 12/6/1967
- 1st engine system full duration altitude test — 1/16/1968
- Design feasibility tests complete — 3/5/1968
- Design validation tests complete — 4/11/1968
- Engine qualification texts complete — 7/11/1968
- Rocketdyne engine contracted awarded — 9/28/1968

Test Program

APOLLO 11

- Design Feasibility Program
  - 12 Injectors
  - 872 Hot fire tests
  - 23,078 seconds total test time
  - 302 combustion stability tests
- Design Verification Program
  - 4 engine systems
  - Environmental tests
  - 209 hot fire tests
  - 5,706 seconds total test time
- Qualification Testing
  - 6 engine systems
  - 308 hot fire tests
  - 14,787 seconds total test time

Appendix I

Program Issues

- Combustion Instability
  - Solution: Injector baffle and acoustic cavity design
  - Stable when bomb tested
  - Validated when cavities blocked and went unstable when bomb tested
- Chamber Erosion
  - Characteristic of injector manifolds
  - Repeatable erosion zones
  - Solution: Unsymmetrical fuel cooling on chamber wall
- Hard engine start (overpressure at start)
  - Fuel entered chamber first and detonated
  - Solution: Added fuel duct volume assuring oxidizer enters chamber first

Legacy Programs Lessons Learned

- Stability was an industry wide issue
- Earlier small research engine instability problem pointed to unique solution
  - 3% of pulse tests showed lower performance
  - Not a detrimental operational problem but not understood
  - Design included a gap between injector and thrust chamber
  - Trapped fuel cause?...increased gap to allow circulation
  - Instability eliminated
- Hindsight analysis revealed gap interrupted combustion instability…acoustic cavity became engine stability aid
- Lesson learned incorporated in ascent engine program

Stability solution key to program success

176 Remembering the Giants

Ascent Engine Design Resolved Issues

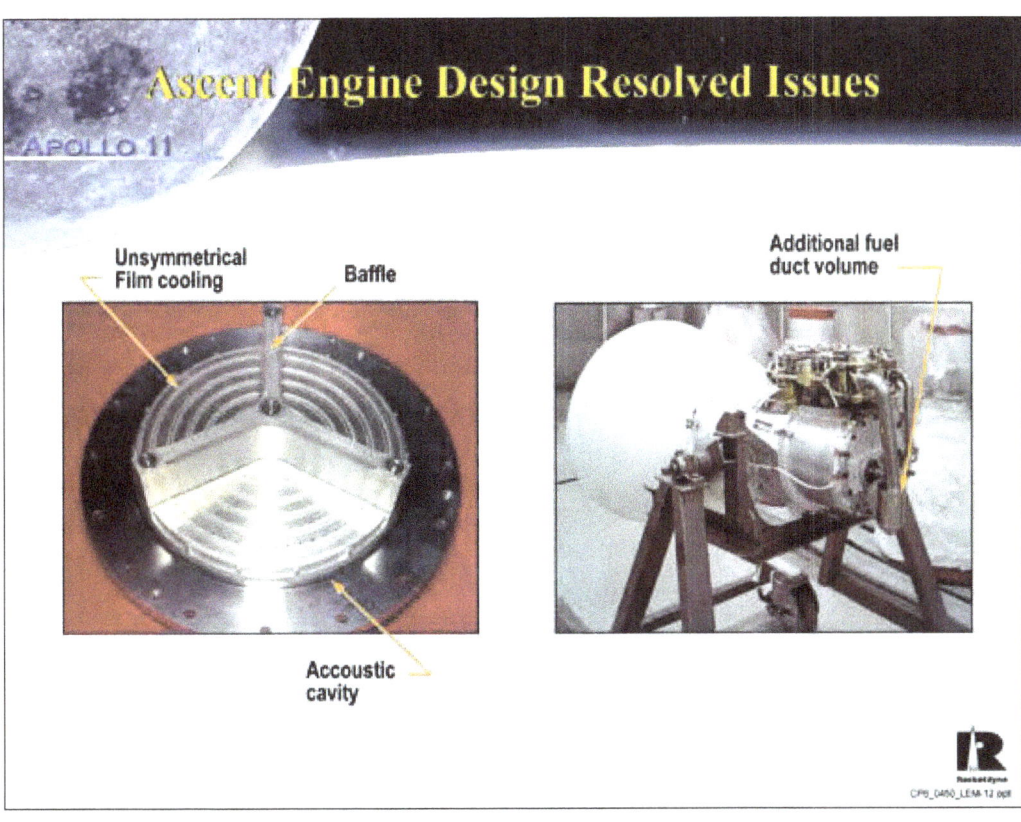

- Unsymmetrical Film cooling
- Baffle
- Accoustic cavity
- Additional fuel duct volume

Development Program Success

- Engine was on critical path
    - Schedule was king!
    - 24/7 test operation, 2 shift manufacturing operation
- Development team included design, development, manufacturing, quality and test staff
- Program communication rapid and effective
- Decision/responsibility flowed down lowest staff level
- Close working relationship with NASA and the vehicle contractor Grumman Aircraft Corp

Appendix I

LEM Engine Program Summary
APOLLO 11

- Unprecedented industry response to NASA challenge
    - Testing initiated one month after award
    - Feasibility testing completed in 8 months (solution demonstrated)
- Prototype Integrated Product and Process Team
    - Totally committed team
    - Process and goal focused
    - All disciplines involved
    - Effective communication

Rocketdyne LEM Engine Legacy: "Can Do" has been earned

Appendix J

Event Photos and Highlights
April 25, 2006

Dr. Shamim Rahman of NASA Stennis Space Center opens the event by introducing NASA SSC Center Director Dr. Richard Gilbrech.

NASA SSC Center Director Dr. Richard Gilbrech welcomes attendees mostly from three NASA Centers (Stennis, Marshall, and Johnson).

Dr. Shamim Rahman of NASA and Mr. Steve Fisher of Rocketdyne introduce the Apollo propulsion speakers.

Appendix J

Mr. Fisher describes the background of this event.

Apollo propulsion speakers accompanied by their wives.

Mr. Bob Biggs describes the development of the Apollo F-1 booster engine.

Mr. Bob Biggs describes the start sequence of the Apollo F-1 booster engine.

Event Photos and Highlights

Mr. Bob Biggs takes audience questions.

Mr. Pete Rodriguez, Director of NASA Marshall Test Lab, poses a question.

Mr. Paul Coffman, engineer on Saturn V J-2 engine, enjoys some of the audience interactions.

Mr. Paul Coffman discusses development challenges on the J-2 engine.

Appendix J

Audience members taking in the lessons from Apollo.

Mr. Paul Coffman takes audience questions.

Mr. Paul Coffman addresses questions.

Mr. Preston Jones, Manager of NASA Marshall Propulsion Department, poses a question on the J-2 engine.

Event Photos and Highlights

Mr. Bob Biggs, developer of the F-1 engine.

Mr. Jerry Pfeifer discusses development of Apollo hypergolic reaction control thrusters.

Mr. Jerry Pfeifer discusses Apollo reaction control thrusters variants.

NASA Stennis Test Complex engineers absorb lessons learned.

Appendix J

Mr. Jerry Pfeifer takes audience questions.

Mr. Tim Harmon discusses his team's propulsion contributions at Rocketdyne during Apollo.

Mr. Tim Harmon shares with audience a picture of a "younger Tim."

Mr. Tim Harmon discusses his team's propulsion contributions at Rocketdyne during Apollo.

Event Photos and Highlights

Mr. Clay Boyce of Aerojet is introduced to begin his discussion of the Apollo service module engine.

Mr. Clay Boyce, developer of the Apollo service propulsion system.

Mr. Clay Boyce discusses the engine value assembly.

Test engineer poses a question.

Apollo Rocket Propulsion Development **185**

Appendix J

Mr. Clay Boyce takes audience questions on Apollo service module propulsion.

Mr. Jerry Elverum discusses the Apollo lunar lander engine.

Mr. Robert Cort of NASA White Sands Test Facility attends as WSTF Propulsion Test Office chief.

Mr. Jerry Elverum, developer of the lunar lander engine.

Event Photos and Highlights

Mr. Miguel Rodriguez (left), Director of NASA SSC Engineering & Science Directorate, and his Deputy Randy Galloway (right) participate in the session along with engineer Jim Ryan (center).

Mr. Bill Arceneaux, center, head of Constellation Program Test and Verification Office.

Mr. Jerry Elverum describes the engine flow control valve and throttling ability.

Mr. Jerry Elverum takes audience questions.

Appendix J

Mr. Tim Harmon gives a second presentation, this time on the lunar ascent engine.

Mr. Tim Harmon takes questions.

Audience members at the closing of the presentations.

Audience upon closing of the lecture that day.

Event Photos and Highlights

Shamim Rahman and wife, Shaheen, at luncheon.

Several seminar presenters: Coffman, Elverum, Biggs and Boyce with their wives at the luncheon.

Tim Harmon with members of NASA SSC Senior Management at Von Braun Tower luncheon.

Further Reading

R. Bilstein, Stages to Saturn: A Technological History of the Apollo/Saturn Launch Vehicles, NASA SP-4206, 1996 (out of print), now published by University of Florida Press, 2003.

R. S. Kraemer, and V. Wheelock, Rocketdyne: Powering Humans into Space (AIAA Education), American Institute of Aeronautics and Astronautics, 2006.

A. Lawrie, Saturn 1/1B: The Complete Manufacturing and Test Records, Apogee Books, 2002.

J. L. Sloop, Liquid Hydrogen as a Propulsion Fuel, 1945-1959. NASA SP-4404, 1978. Out of print.

G. P. Sutton, History of Liquid Propellant Rocket Engines, American Institute of Aeronautics and Astronautics, 2006.

G. P. Sutton and Biblarz, Rocket Propulsion Elements, 7th Edition, John Wiley and Sons, 2001.

A. Young, The Saturn V F-1 Engine: Powering Apollo into History, Springer Praxis, 2008.

Saturn V, AS-501 Vehicle
(NASA Image number: S67-50531)

Glossary

Ablative – Material that evaporates or vaporizes.

Ablestar – The Thor-Ablestar, or Thor AbleStar, was an American expendable launch system consisting of a Thor missile, with an Ablestar-powered upper stage.

Actuate – To put into motion or action.

Aerozine-50 – A 50/50 mix of hydrazine and unsymmetrical dimethylhydrazine, it was developed in the late 1950s as a storable, high-energy, hypergolic rocket fuel.

APG – Adverse pressure gradient.

Attitude – The orientation of a spacecraft relative to its direction of motion.

Cf - Thrust coefficient; a measure of the amplification of thrust due to gas expansion in a particular nozzle as compared with the thrust that would be exerted if the chamber pressure acted only over the throat area of the nozzle.

Calrod® – An elongated electrical heater unit, much like the coiled burner/heating element on an electric stove. The trademark name has become a generic moniker for any electric heater in the shape of a straight rod.

Cavitation – The formation of vapor bubbles of a flowing liquid in a region where the pressure of the liquid falls below its vapor pressure. Mechanical forces - such as the moving blades of a ship's propeller - can cause this.

Chlorine trifluoride (ClF_3) – A colorless, poisonous, corrosive and very reactive gas that is primarily of interest as a component in rocket fuels; in industrial cleaning and etching operations in the semiconductor industry; in nuclear reactor fuel processing; and in other industrial operations.

Columbium – Also known as niobium, this chemical element is used as an alloying element in steels and superalloys.

Deep throttling – Refers to drastically reducing the thrust of an engine by restricting flow of propellants to less than 100 percent of capacity. The challenge is to do so without creating instability in the engine. Most engines are designed to deliver constant thrust. Many can be throttled down somewhat. For instance, the space shuttle main engine can be throttled down to 60 percent thrust without instability occurring. However, to design an engine that can be throttled down to 10 percent or less of optimal thrust is very challenging.

Delta-P – Delta pressure refers to the difference between standard atmospheric air pressure and the air pressure read by a gauge or test equipment.

DIGR – Datelink Independent Gateway Retrofit

Glossary

F-1 – An engine developed by Rocketdyne and used to power the Saturn V vehicle. Five F-1 engines were used in the S-IC first stage of each Saturn V, which served as the main launch vehicle in the Apollo program. The F-1 is still the most powerful liquid-fueled rocket engine developed.

Grumman – The Grumman Aircraft Engineering Corp. was founded in 1929 and grew into a leading twentieth-century American producer of military and civilian aircraft. It later was renamed the Grumman Aerospace Corp.. In 1994, it was acquired by Northrop Corp. to form Northrop Grumman.

Helmholtz resonator – Named for German physician and physicist Hermann Ludwig Ferdinand von Helmholtz, it is a container of gas (usually air) with an open hole (or neck or port). A volume of air in and near the open hole vibrates because of the "springiness" of the air inside. A common example is an empty bottle: the air inside vibrates when one blows across the top. Source: www.phys.unsw.edu.au/jw/Helmholtz.html

Hypergolic – Refers to an ignition process that involves a rocket propellant mixing with an oxidizer and igniting upon contact. No ignition spark is needed. In the process, a hypergolic fluid is hermetically sealed in a cartridge. As long as it is sealed, the fluid is stable. However, the fluid spontaneously ignites when it comes into contact with any form of oxygen. In starting an engine, increased fuel pressure within the igniter fuel system ruptures the cartridge. Fuel and the hypergolic fluid enter the chamber via separate lines. When the hypergolic fluid comes in contact with an oxidizer already present, it ignites, initiating fuel combustion.

Hysteresis – A lag of effect when the forces acting on a body are changed

IPPD – Integrated Product and Process Development

IR&D – Independent Research and Development

ISP (specific impulse) – A measure of the "fuel efficiency" of a rocket or jet engine. It represents the impulse (change in momentum) per unit of propellant. The higher the specific impulse, the less propellant is needed to gain a given amount of momentum.Specific impulse is a useful value to compare engines, much like "miles per gallon" for cars. A propulsion method with a higher specific impulse is more propellant-efficient.

J-2 – An engine used for the second and third stages of the Saturn V moon rocket. The J-2 engine used liquid hydrogen and liquid oxygen propellants for a maximum thrust of 225,000 pounds. It was the first manned booster engine to use liquid hydrogen as a propellant and the first large booster engine designed to be restarted multiple times during a mission.

LEM – Lunar Excursion Module

lbf – Pounds of force

Mach – The speed of an object moving through air, divided by the speed of sound. Thus, something traveling Mach 7 is traveling seven times the speed of sound. According to the aerospaceweb.org Web site, the generally accepted speed of sound is 1,116.4 feet per second (or 761.2 miles per hour)

Mandrel – A spindle or axle used to support material being machined or milled; a core around which materials may be cast and shaped, or a shaft on which a working tool is mounted

Man-rated – A term used to signify that a spacecraft, launch vehicle or airplane is worthy of transporting humans. Referred to as "human-rated" in most instances now.

MMH – Monomethylhydrazine, a volatile hydrazine chemical used as a fuel in bi-propellant rocket engines and frequently in hypergolic mixtures.

Molybdenum – A silvery metal that has the sixth-highest melting point of any element. It readily forms hard, stable carbides and often is used in high-strength steel alloys.

N_2O_4 – Dinitrogen tetroxide (nitrogen tetroxide or nitrogen peroxide). It is a powerful oxidizer, highly toxic and corrosive. N_2O_4 is hypergolic with various forms of hydrazine, i.e., they burn on contact without a separate ignition source, making them popular bi-propellant rocket fuels

North American Aviation Inc. – Founded in 1928, North American was a major American aircraft manufacturer, responsible for such crafts as the P-51 Mustang fighter, the F-86 Savre jet fighter, the Apollo Command and Service modules and the space shuttle orbiter. Through various mergers, it now is part of The Boeing Company.

OME – Orbital maneuvering engine

Pintle – A a pin or bolt used as part of a pivot or hinge. In rocketry, a pintle engine uses a single-feed fuel injector rather than the hundreds of smaller holes used in a typical rocket engine. That lowers the cost of engine manufacture while surrendering some performance. The concept originally was developed by the Grumman Aerospace Corp. for the Apollo Lunar Module.

PFRT – Preliminary Flight Rating Test

Psi – Pounds per square inch (measure of pressure)

Reaction Motors Inc. – Formed in 1941 by four members of the American Rocket Society, the company developed the rocket engines used by the early X planes, including the Bell X-1 that first broke the sound barrier in 1947 and the famous North American X-15. The company was taken over by Thiokol Chemical in 1958 to become the Reaction Motor Division. Later, the division became Thiokol Propulsion, which was acquired by ATK (Alliant Techsystems) in 2001.

Retort – A closed laboratory vessel with an outlet tube, used for distillation, sublimation, or decomposition of substances by heat.

RFNA – Red fuming nitric acid, a storable oxidizer used as a rocket propellant. It consists mainly of nitric acid, also containing dinitrogen tetroxide and water.

RP-1 – Rocket propellant 1, a highly refined form of kerosene used as a rocket fuel, usually with liquid oxygen as an oxidizer.

SSME – Space shuttle main engine

Scarfed nozzle – A two-dimensional, non-symmetric nozzle with a flat lower wall (cowl) that is truncated to save weight.

SPS – Service Propulsion System. In the Apollo Program, it was used to place spacecraft into and out of lunar orbit, and for mid-course corrections between the Earth and the moon. The Apollo SPS used a single AJ10-137 engine.

Theater High Altitude Area Defense (THAAD) – Formerly known as the Terminal High Altitude Area Defense, ut is a U.S. Army project to develop a system to shoot down short- and medium-range ballistic missiles.

TRW Inc. – a company active in the early development of missile systems and spacecraft, most notably the NASA deep space satellites Pioneer 10 and 11. TRW also pioneered systems engineering. The 1958 merger of Thompson Products Co. with the Ramo-Wooldridge Corp. was named Thompson Ramo Wooldridge Inc., then shortened to TRW Inc. in 1965.

Unsymmetrical dimethylhydrazine (UDMH) – A toxic, volatile, clear liquid that turns yellowish on exposure to air and absorbs oxygen and carbon dioxide. It is also hygroscopic, meaning it attracts water molecules from the surrounding environment. It mixes completely with water, ethanol, and kerosene. In concentrations between 2.5 percent and 95 percent in air, its vapors are flammable. It often is used in hypergolic rocket fuels because it is stable and can be kept loaded in rocket fuel systems for long periods. UDMH functions as a starter fuel to start combustion and warm the rocket engine prior to switching to kerosene.

Venturi – A tube with a tapered constriction that causes an increase in the velocity of flow of a fluid and a corresponding decrease in fluid pressure; a specially shaped tube that affects the speed and pressure of a flowing fluid

The NASA History Series

REFERENCE WORKS, NASA SP-4000:

Grimwood, James M. *Project Mercury: A Chronology*. NASA SP-4001, 1963.

Grimwood, James M., and Barton C. Hacker, with Peter J. Vorzimmer. *Project Gemini Technology and Operations: A Chronology*. NASA SP-4002, 1969.

Link, Mae Mills. *Space Medicine in Project Mercury*. NASA SP-4003, 1965.

Astronautics and Aeronautics, 1963: Chronology of Science, Technology, and Policy. NASA SP-4004, 1964.

Astronautics and Aeronautics, 1964: Chronology of Science, Technology, and Policy. NASA SP-4005, 1965.

Astronautics and Aeronautics, 1965: Chronology of Science, Technology, and Policy. NASA SP-4006, 1966.

Astronautics and Aeronautics, 1966: Chronology of Science, Technology, and Policy. NASA SP-4007, 1967.

Astronautics and Aeronautics, 1967: Chronology of Science, Technology, and Policy. NASA SP-4008, 1968.

Ertel, Ivan D., and Mary Louise Morse. *The Apollo Spacecraft: A Chronology, Volume I, Through November 7, 1962*. NASA SP-4009, 1969.

Morse, Mary Louise, and Jean Kernahan Bays. *The Apollo Spacecraft: A Chronology, Volume II, November 8, 1962–September 30, 1964*. NASA SP-4009, 1973.

Brooks, Courtney G., and Ivan D. Ertel. *The Apollo Spacecraft: A Chronology, Volume III, October 1, 1964–January 20, 1966*. NASA SP-4009, 1973.

Ertel, Ivan D., and Roland W. Newkirk, with Courtney G. Brooks. *The Apollo Spacecraft: A Chronology, Volume IV, January 21, 1966–July 13, 1974*. NASA SP-4009, 1978.

Astronautics and Aeronautics, 1968: Chronology of Science, Technology, and Policy. NASA SP-4010, 1969.

Newkirk, Roland W., and Ivan D. Ertel, with Courtney G. Brooks. *Skylab: A Chronology*. NASA SP-4011, 1977.

Van Nimmen, Jane, and Leonard C. Bruno, with Robert L. Rosholt. *NASA Historical Data Book, Vol. I: NASA Resources, 1958–1968*. NASA SP-4012, 1976, rep. ed. 1988.

Ezell, Linda Neuman. *NASA Historical Data Book, Vol. II: Programs and Projects, 1958–1968*. NASA SP-4012, 1988.

Ezell, Linda Neuman. *NASA Historical Data Book, Vol. III: Programs and Projects, 1969–1978*. NASA SP-4012, 1988.

Gawdiak, Ihor, with Helen Fedor. *NASA Historical Data Book, Vol. IV: NASA Resources, 1969–1978*. NASA SP-4012, 1994.

Rumerman, Judy A. *NASA Historical Data Book, Vol. V: NASA Launch Systems, Space Transportation, Human Spaceflight, and Space Science, 1979–1988*. NASA SP-4012, 1999.

Rumerman, Judy A. *NASA Historical Data Book, Vol. VI: NASA Space Applications, Aeronautics and Space Research and Technology, Tracking and Data Acquisition/Support Operations, Commercial Programs, and Resources, 1979–1988*. NASA SP-4012, 1999.

Rumerman, Judy A. *NASA Historical Data Book, Vol. VII: NASA Launch Systems, Space Transportation, Human Spaceflight, and Space Science, 1989–1998*. NASA SP-2009-4012.

Astronautics and Aeronautics, 1969: Chronology of Science, Technology, and Policy. NASA SP-4014, 1970.

Astronautics and Aeronautics, 1970: Chronology of Science, Technology, and Policy. NASA SP-4015, 1972.

Astronautics and Aeronautics, 1971: Chronology of Science, Technology, and Policy. NASA SP-4016, 1972.

Astronautics and Aeronautics, 1972: Chronology of Science, Technology, and Policy. NASA SP-4017, 1974.

Astronautics and Aeronautics, 1973: Chronology of Science, Technology, and Policy. NASA SP-4018, 1975.

Astronautics and Aeronautics, 1974: Chronology of Science, Technology, and Policy. NASA SP-4019, 1977.

Astronautics and Aeronautics, 1975: Chronology of Science, Technology, and Policy. NASA SP-4020, 1979.

Astronautics and Aeronautics, 1976: Chronology of Science, Technology, and Policy. NASA SP-4021, 1984.

Astronautics and Aeronautics, 1977: Chronology of Science, Technology, and Policy. NASA SP-4022, 1986.

Astronautics and Aeronautics, 1978: Chronology of Science, Technology, and Policy. NASA SP-4023, 1986.

Astronautics and Aeronautics, 1979–1984: Chronology of Science, Technology, and Policy. NASA SP-4024, 1988.

Astronautics and Aeronautics, 1985: Chronology of Science, Technology, and Policy. NASA SP-4025, 1990.

Noordung, Hermann. *The Problem of Space Travel: The Rocket Motor*. Edited by Ernst Stuhlinger and J.D. Hunley, with Jennifer Garland. NASA SP-4026, 1995.

Astronautics and Aeronautics, 1986–1990: A Chronology. NASA SP-4027, 1997.

Astronautics and Aeronautics, 1991–1995: A Chronology. NASA SP-2000-4028, 2000.

Orloff, Richard W. *Apollo by the Numbers: A Statistical Reference.* NASA SP-2000-4029, 2000.

Lewis, Marieke and Swanson, Ryan. *Aeronautics and Astronautics: A Chronology, 1996-2000.* NASA SP-2009-4030, 2009.

MANAGEMENT HISTORIES, NASA SP-4100:

Rosholt, Robert L. *An Administrative History of NASA, 1958–1963.* NASA SP-4101, 1966.

Levine, Arnold S. *Managing NASA in the Apollo Era.* NASA SP-4102, 1982.

Roland, Alex. *Model Research: The National Advisory Committee for Aeronautics, 1915–1958.* NASA SP-4103, 1985.

Fries, Sylvia D. *NASA Engineers and the Age of Apollo.* NASA SP-4104, 1992.

Glennan, T. Keith. *The Birth of NASA: The Diary of T. Keith Glennan.* Edited by J.D. Hunley. NASA SP-4105, 1993.

Seamans, Robert C. *Aiming at Targets: The Autobiography of Robert C. Seamans.* NASA SP-4106, 1996.

Garber, Stephen J., editor. *Looking Backward, Looking Forward: Forty Years of Human Spaceflight Symposium.* NASA SP-2002-4107, 2002.

Mallick, Donald L. with Peter W. Merlin. *The Smell of Kerosene: A Test Pilot's Odyssey.* NASA SP-4108, 2003.

Iliff, Kenneth W. and Curtis L. Peebles. *From Runway to Orbit: Reflections of a NASA Engineer.* NASA SP-2004-4109, 2004.

Chertok, Boris. *Rockets and People, Volume 1.* NASA SP-2005-4110, 2005.

Chertok, Boris. *Rockets and People: Creating a Rocket Industry, Volume II.* NASA SP-2006-4110, 2006.

Laufer, Alexander, Todd Post, and Edward Hoffman. *Shared Voyage: Learning and Unlearning from Remarkable Projects.* NASA SP-2005-4111, 2005.

Dawson, Virginia P., and Mark D. Bowles. *Realizing the Dream of Flight: Biographical Essays in Honor of the Centennial of Flight, 1903–2003.* NASA SP-2005-4112, 2005.

Mudgway, Douglas J. *William H. Pickering: America's Deep Space Pioneer.* NASA SP-2008-4113.

PROJECT HISTORIES, NASA SP-4200:

Swenson, Loyd S., Jr., James M. Grimwood, and Charles C. Alexander. *This New Ocean: A History of Project Mercury.* NASA SP-4201, 1966; reprinted 1999.

NASA History Series

Green, Constance McLaughlin, and Milton Lomask. *Vanguard: A History*. NASA SP-4202, 1970; rep. ed. Smithsonian Institution Press, 1971.

Hacker, Barton C., and James M. Grimwood. *On Shoulders of Titans: A History of Project Gemini*. NASA SP-4203, 1977, reprinted 2002.

Benson, Charles D., and William Barnaby Faherty. *Moonport: A History of Apollo Launch Facilities and Operations*. NASA SP-4204, 1978.

Brooks, Courtney G., James M. Grimwood, and Loyd S. Swenson, Jr. *Chariots for Apollo: A History of Manned Lunar Spacecraft*. NASA SP-4205, 1979.

Bilstein, Roger E. *Stages to Saturn: A Technological History of the Apollo/Saturn Launch Vehicles*. NASA SP-4206, 1980 and 1996.

Compton, W. David, and Charles D. Benson. *Living and Working in Space: A History of Skylab*. NASA SP-4208, 1983.

Ezell, Edward Clinton, and Linda Neuman Ezell. *The Partnership: A History of the Apollo-Soyuz Test Project*. NASA SP-4209, 1978.

Hall, R. Cargill. *Lunar Impact: A History of Project Ranger*. NASA SP-4210, 1977.

Newell, Homer E. *Beyond the Atmosphere: Early Years of Space Science*. NASA SP-4211, 1980.

Ezell, Edward Clinton, and Linda Neuman Ezell. *On Mars: Exploration of the Red Planet, 1958–1978*. NASA SP-4212, 1984.

Pitts, John A. *The Human Factor: Biomedicine in the Manned Space Program to 1980*. NASA SP-4213, 1985.

Compton, W. David. *Where No Man Has Gone Before: A History of Apollo Lunar Exploration Missions*. NASA SP-4214, 1989.

Naugle, John E. *First Among Equals: The Selection of NASA Space Science Experiments*. NASA SP-4215, 1991.

Wallace, Lane E. *Airborne Trailblazer: Two Decades with NASA Langley's 737 Flying Laboratory*. NASA SP-4216, 1994.

Butrica, Andrew J., ed. *Beyond the Ionosphere: Fifty Years of Satellite Communications*. NASA SP-4217, 1997.

Butrica, Andrew J. *To See the Unseen: A History of Planetary Radar Astronomy*. NASA SP-4218, 1996.

Mack, Pamela E., ed. *From Engineering Science to Big Science: The NACA and NASA Collier Trophy Research Project Winners*. NASA SP-4219, 1998.

Reed, R. Dale. *Wingless Flight: The Lifting Body Story*. NASA SP-4220, 1998.

Heppenheimer, T. A. *The Space Shuttle Decision: NASA's Search for a Reusable Space Vehicle*. NASA SP-4221, 1999.

Hunley, J. D., ed. *Toward Mach 2: The Douglas D-558 Program*. NASA SP-4222, 1999.

Swanson, Glen E., ed. *"Before This Decade is Out . . ." Personal Reflections on the Apollo Program*. NASA SP-4223, 1999.

Tomayko, James E. *Computers Take Flight: A History of NASA's Pioneering Digital Fly-By-Wire Project*. NASA SP-4224, 2000.

Morgan, Clay. *Shuttle-Mir: The United States and Russia Share History's Highest Stage*. NASA SP-2001-4225.

Leary, William M. *"We Freeze to Please:" A History of NASA's Icing Research Tunnel and the Quest for Safety*. NASA SP-2002-4226, 2002.

Mudgway, Douglas J. *Uplink-Downlink: A History of the Deep Space Network, 1957–1997*. NASA SP-2001-4227.

Dawson, Virginia P., and Mark D. Bowles. *Taming Liquid Hydrogen: The Centaur Upper Stage Rocket, 1958–2002*. NASA SP-2004-4230.

Meltzer, Michael. *Mission to Jupiter: A History of the Galileo Project*. NASA SP-2007-4231.

Heppenheimer, T. A. *Facing the Heat Barrier: A History of Hypersonics*. NASA SP-2007-4232.

Tsiao, Sunny. *"Read You Loud and Clear!" The Story of NASA's Spaceflight Tracking and Data Network*. NASA SP-2007-4233.

CENTER HISTORIES, NASA SP-4300:

Rosenthal, Alfred. *Venture into Space: Early Years of Goddard Space Flight Center*. NASA SP-4301, 1985.

Hartman, Edwin, P. *Adventures in Research: A History of Ames Research Center, 1940–1965*. NASA SP-4302, 1970.

Hallion, Richard P. *On the Frontier: Flight Research at Dryden, 1946–1981*. NASA SP-4303, 1984.

Muenger, Elizabeth A. *Searching the Horizon: A History of Ames Research Center, 1940–1976*. NASA SP-4304, 1985.

Hansen, James R. *Engineer in Charge: A History of the Langley Aeronautical Laboratory, 1917–1958*. NASA SP-4305, 1987.

Dawson, Virginia P. *Engines and Innovation: Lewis Laboratory and American Propulsion Technology*. NASA SP-4306, 1991.

Dethloff, Henry C. *"Suddenly Tomorrow Came . . .": A History of the Johnson Space Center, 1957–1990*. NASA SP-4307, 1993.

Hansen, James R. *Spaceflight Revolution: NASA Langley Research Center from Sputnik to Apollo*. NASA SP-4308, 1995.

Wallace, Lane E. *Flights of Discovery: An Illustrated History of the Dryden Flight Research Center*. NASA SP-4309, 1996.

Herring, Mack R. *Way Station to Space: A History of the John C. Stennis Space Center*. NASA SP-4310, 1997.

Wallace, Harold D., Jr. *Wallops Station and the Creation of an American Space Program*. NASA SP-4311, 1997.

Wallace, Lane E. *Dreams, Hopes, Realities. NASA's Goddard Space Flight Center: The First Forty Years*. NASA SP-4312, 1999.

Dunar, Andrew J., and Stephen P. Waring. *Power to Explore: A History of Marshall Space Flight Center, 1960–1990*. NASA SP-4313, 1999.

Bugos, Glenn E. *Atmosphere of Freedom: Sixty Years at the NASA Ames Research Center*. NASA SP-2000-4314, 2000.

Schultz, James. *Crafting Flight: Aircraft Pioneers and the Contributions of the Men and Women of NASA Langley Research Center*. NASA SP-2003-4316, 2003.

Bowles, Mark D. *Science in Flux: NASA's Nuclear Program at Plum Brook Station, 1955–2005*. NASA SP-2006-4317.

Wallace, Lane E. *Flights of Discovery: An Illustrated History of the Dryden Flight Research Center*. NASA SP-4318, 2007. Revised version of SP-4309.

GENERAL HISTORIES, NASA SP-4400:

Corliss, William R. *NASA Sounding Rockets, 1958–1968: A Historical Summary*. NASA SP-4401, 1971.

Wells, Helen T., Susan H. Whiteley, and Carrie Karegeannes. *Origins of NASA Names*. NASA SP-4402, 1976.

Anderson, Frank W., Jr. *Orders of Magnitude: A History of NACA and NASA, 1915–1980*. NASA SP-4403, 1981.

Sloop, John L. *Liquid Hydrogen as a Propulsion Fuel, 1945–1959*. NASA SP-4404, 1978.

Roland, Alex. *A Spacefaring People: Perspectives on Early Spaceflight*. NASA SP-4405, 1985.

Bilstein, Roger E. *Orders of Magnitude: A History of the NACA and NASA, 1915–1990*. NASA SP-4406, 1989.

Logsdon, John M., ed., with Linda J. Lear, Jannelle Warren Findley, Ray A. Williamson, and Dwayne A. Day. *Exploring the Unknown: Selected Documents in the History of the U.S. Civil Space Program, Volume I, Organizing for Exploration*. NASA SP-4407, 1995.

Logsdon, John M., ed, with Dwayne A. Day, and Roger D. Launius. *Exploring the Unknown: Selected Documents in the History of the U.S. Civil Space Program, Volume II, External Relationships.* NASA SP-4407, 1996.

Logsdon, John M., ed., with Roger D. Launius, David H. Onkst, and Stephen J. Garber. *Exploring the Unknown: Selected Documents in the History of the U.S. Civil Space Program, Volume III, Using Space.* NASA SP-4407, 1998.

Logsdon, John M., ed., with Ray A. Williamson, Roger D. Launius, Russell J. Acker, Stephen J. Garber, and Jonathan L. Friedman. *Exploring the Unknown: Selected Documents in the History of the U.S. Civil Space Program, Volume IV, Accessing Space.* NASA SP-4407, 1999.

Logsdon, John M., ed., with Amy Paige Snyder, Roger D. Launius, Stephen J. Garber, and Regan Anne Newport. *Exploring the Unknown: Selected Documents in the History of the U.S. Civil Space Program, Volume V, Exploring the Cosmos.* NASA SP-4407, 2001.

Logsdon, John M., ed., with Stephen J. Garber, Roger D. Launius, and Ray A. Williamson. *Exploring the Unknown: Selected Documents in the History of the U.S. Civil Space Program, Volume VI: Space and Earth Science.* NASA SP-2004-4407, 2004.

Logsdon, John M., ed., with Roger D. Launius. *Exploring the Unknown: Selected Documents in the History of the U.S. Civil Space Program, Volume VII: Human Spaceflight: Projects Mercury, Gemini, and Apollo.* NASA SP-2008-4407, 2008.

Siddiqi, Asif A., *Challenge to Apollo: The Soviet Union and the Space Race, 1945–1974.* NASA SP-2000-4408, 2000.

Hansen, James R., ed. *The Wind and Beyond: Journey into the History of Aerodynamics in America, Volume 1, The Ascent of the Airplane.* NASA SP-2003-4409, 2003.

Hansen, James R., ed. *The Wind and Beyond: Journey into the History of Aerodynamics in America, Volume 2, Reinventing the Airplane.* NASA SP-2007-4409, 2007.

Hogan, Thor. *Mars Wars: The Rise and Fall of the Space Exploration Initiative.* NASA SP-2007-4410, 2007.

MONOGRAPHS IN AEROSPACE HISTORY (SP-4500 SERIES):

Launius, Roger D., and Aaron K. Gillette, compilers. *Toward a History of the Space Shuttle: An Annotated Bibliography.* Monograph in Aerospace History, No. 1, 1992.

Launius, Roger D., and J. D. Hunley, compilers. *An Annotated Bibliography of the Apollo Program.* Monograph in Aerospace History No. 2, 1994.

Launius, Roger D. *Apollo: A Retrospective Analysis.* Monograph in Aerospace History, No. 3, 1994.

Hansen, James R. *Enchanted Rendezvous: John C. Houbolt and the Genesis of the Lunar-Orbit Rendezvous Concept.* Monograph in Aerospace History, No. 4, 1995.

Gorn, Michael H. *Hugh L. Dryden's Career in Aviation and Space*. Monograph in Aerospace History, No. 5, 1996.

Powers, Sheryll Goecke. *Women in Flight Research at NASA Dryden Flight Research Center from 1946 to 1995*. Monograph in Aerospace History, No. 6, 1997.

Portree, David S. F., and Robert C. Trevino. *Walking to Olympus: An EVA Chronology*. Monograph in Aerospace History, No. 7, 1997.

Logsdon, John M., moderator. *Legislative Origins of the National Aeronautics and Space Act of 1958: Proceedings of an Oral History Workshop*. Monograph in Aerospace History, No. 8, 1998.

Rumerman, Judy A., compiler. *U.S. Human Spaceflight, A Record of Achievement 1961–1998*. Monograph in Aerospace History, No. 9, 1998.

Portree, David S. F. *NASA's Origins and the Dawn of the Space Age*. Monograph in Aerospace History, No. 10, 1998.

Logsdon, John M. *Together in Orbit: The Origins of International Cooperation in the Space Station*. Monograph in Aerospace History, No. 11, 1998.

Phillips, W. Hewitt. *Journey in Aeronautical Research: A Career at NASA Langley Research Center*. Monograph in Aerospace History, No. 12, 1998.

Braslow, Albert L. *A History of Suction-Type Laminar-Flow Control with Emphasis on Flight Research*. Monograph in Aerospace History, No. 13, 1999.

Logsdon, John M., moderator. *Managing the Moon Program: Lessons Learned From Apollo*. Monograph in Aerospace History, No. 14, 1999.

Perminov, V. G. *The Difficult Road to Mars: A Brief History of Mars Exploration in the Soviet Union*. Monograph in Aerospace History, No. 15, 1999.

Tucker, Tom. *Touchdown: The Development of Propulsion Controlled Aircraft at NASA Dryden*. Monograph in Aerospace History, No. 16, 1999.

Maisel, Martin, Demo J. Giulanetti, and Daniel C. Dugan. *The History of the XV-15 Tilt Rotor Research Aircraft: From Concept to Flight*. Monograph in Aerospace History, No. 17, 2000. NASA SP-2000-4517.

Jenkins, Dennis R. *Hypersonics Before the Shuttle: A Concise History of the X-15 Research Airplane*. Monograph in Aerospace History, No. 18, 2000. NASA SP-2000-4518.

Chambers, Joseph R. *Partners in Freedom: Contributions of the Langley Research Center to U.S. Military Aircraft of the 1990s*. Monograph in Aerospace History, No. 19, 2000. NASA SP-2000-4519.

Waltman, Gene L. *Black Magic and Gremlins: Analog Flight Simulations at NASA's Flight Research Center*. Monograph in Aerospace History, No. 20, 2000. NASA SP-2000-4520.

NASA History Series

Portree, David S. F. *Humans to Mars: Fifty Years of Mission Planning, 1950–2000.* Monograph in Aerospace History, No. 21, 2001. NASA SP-2001-4521.

Thompson, Milton O., with J. D. Hunley. *Flight Research: Problems Encountered and What they Should Teach Us.* Monograph in Aerospace History, No. 22, 2001. NASA SP-2001-4522.

Tucker, Tom. *The Eclipse Project.* Monograph in Aerospace History, No. 23, 2001. NASA SP-2001-4523.

Siddiqi, Asif A. *Deep Space Chronicle: A Chronology of Deep Space and Planetary Probes 1958–2000.* Monograph in Aerospace History, No. 24, 2002. NASA SP-2002-4524.

Merlin, Peter W. *Mach 3+: NASA/USAF YF-12 Flight Research, 1969–1979.* Monograph in Aerospace History, No. 25, 2001. NASA SP-2001-4525.

Anderson, Seth B. *Memoirs of an Aeronautical Engineer: Flight Tests at Ames Research Center: 1940–1970.* Monograph in Aerospace History, No. 26, 2002. NASA SP-2002-4526.

Renstrom, Arthur G. *Wilbur and Orville Wright: A Bibliography Commemorating the One-Hundredth Anniversary of the First Powered Flight on December 17, 1903.* Monograph in Aerospace History, No. 27, 2002. NASA SP-2002-4527.

No monograph 28.

Chambers, Joseph R. *Concept to Reality: Contributions of the NASA Langley Research Center to U.S. Civil Aircraft of the 1990s.* Monograph in Aerospace History, No. 29, 2003. SP-2003-4529.

Peebles, Curtis, editor. *The Spoken Word: Recollections of Dryden History, The Early Years.* Monograph in Aerospace History, No. 30, 2003. SP-2003-4530.

Jenkins, Dennis R., Tony Landis, and Jay Miller. *American X-Vehicles: An Inventory- X-1 to X-50.* Monograph in Aerospace History, No. 31, 2003. SP-2003-4531.

Renstrom, Arthur G. *Wilbur and Orville Wright: A Chronology Commemorating the One-Hundredth Anniversary of the First Powered Flight on December 17, 1903.* Monograph in Aerospace History, No. 32, 2003. NASA SP-2003-4532.

Bowles, Mark D., and Robert S. Arrighi. *NASA's Nuclear Frontier: The Plum Brook Research Reactor.* Monograph in Aerospace History, No. 33, 2004. (SP-2004-4533).

Wallace, Lane and Christian Gelzer. *Nose Up: High Angle-of-Attack and Thrust Vectoring Research at NASA Dryden, 1979-2001.* Monograph in Aerospace History No. 34, 2009. NASA SP-2009-4534.

Matranga, Gene J., C. Wayne Ottinger, Calvin R. Jarvis, and D. Christian Gelzer. *Unconventional, Contrary, and Ugly: The Lunar Landing Research Vehicle.* Monograph in Aerospace History, No. 35, 2006. NASA SP-2004-4535.

McCurdy, Howard E. *Low Cost Innovation in Spaceflight: The History of the Near Earth Asteroid Rendezvous (NEAR) Mission.* Monograph in Aerospace History, No. 36, 2005. NASA SP-2005-4536.

Seamans, Robert C., Jr. *Project Apollo: The Tough Decisions.* Monograph in Aerospace History, No. 37, 2005. NASA SP-2005-4537.

Lambright, W. Henry. *NASA and the Environment: The Case of Ozone Depletion.* Monograph in Aerospace History, No. 38, 2005. NASA SP-2005-4538.

Chambers, Joseph R. *Innovation in Flight: Research of the NASA Langley Research Center on Revolutionary Advanced Concepts for Aeronautics.* Monograph in Aerospace History, No. 39, 2005. NASA SP-2005-4539.

Phillips, W. Hewitt. *Journey Into Space Research: Continuation of a Career at NASA Langley Research Center.* Monograph in Aerospace History, No. 40, 2005. NASA SP-2005-4540.

Rumerman, Judy A., Chris Gamble, and Gabriel Okolski, compilers. *U.S. Human Spaceflight: A Record of Achievement, 1961–2006.* Monograph in Aerospace History No. 41, 2007. NASA SP-2007-4541.

Dick, Steven J.; Garber, Stephen J.; and Odom, Jane H. *Research in NASA History.* Monograph in Aerospace History No. 43, 2009. NASA SP-2009-4543.

Merlin, Peter W., *Ikhana: Unmanned Aircraft System Western States Fire Missions.* Monograph in Aerospace History #44. NASA SP-2009-4544.

ELECTRONIC MEDIA (SP-4600 SERIES)

Remembering Apollo 11: The 30th Anniversary Data Archive CD-ROM. NASA SP-4601, 1999.

Remembering Apollo 11: The 35th Anniversary Data Archive CD-ROM. NASA SP-2004-4601, 2004. This is an update of the 1999 edition.

The Mission Transcript Collection: U.S. Human Spaceflight Missions from Mercury Redstone 3 to Apollo 17. SP-2000-4602, 2001.

Shuttle-Mir: the United States and Russia Share History's Highest Stage. NASA SP-2001-4603, 2002.

U.S. Centennial of Flight Commission presents Born of Dreams – Inspired by Freedom. NASA SP-2004-4604, 2004.

Of Ashes and Atoms: A Documentary on the NASA Plum Brook Reactor Facility. NASA SP-2005-4605.

Taming Liquid Hydrogen: The Centaur Upper Stage Rocket Interactive CD-ROM. NASA SP-2004-4606, 2004.

Fueling Space Exploration: The History of NASA's Rocket Engine Test Facility DVD. NASA SP-2005-4607.

Altitude Wind Tunnel at NASA Glenn Research Center: An Interactive History CD-ROM. NASA SP-2008-4608.

CONFERENCE PROCEEDINGS (SP-4700 SERIES)

Dick, Steven J., and Keith Cowing, ed. *Risk and Exploration: Earth, Sea and the Stars.* NASA SP-2005-4701.

Dick, Steven J., and Roger D. Launius. *Critical Issues in the History of Spaceflight.* NASA SP-2006-4702.

Dick, Steven J., ed. *Remembering the Space Age: Proceedings of the 50th Anniversary Conference.* NASA SP-2008-4703.

SOCIETAL IMPACT (SP-4800 SERIES)

Dick, Steven J., and Roger D. Launius. *Societal Impact of Spaceflight.* NASA SP-2007-4801.

www.ingramcontent.com/pod-product-compliance
Lightning Source LLC
Chambersburg PA
CBHW082120230426
43671CB00015B/2753